Light Pollution in Metropolises

Emlyn Etienne Goronczy

Light Pollution in Metropolises

Analysis, Impacts and Solutions

Springer

Emlyn Etienne Goronczy
Hannover, Germany

ISBN 978-3-658-29722-0 ISBN 978-3-658-29723-7 (eBook)
https://doi.org/10.1007/978-3-658-29723-7

The translation was done with the help of artificial intelligence (machine translation by the service DeepL.com). A subsequent human revision was done primarily in terms of content.

This Springer imprint is published by the registered company Springer Fachmedien Wiesbaden GmbH, part of Springer Nature.
The registered company address is: Abraham-Lincoln-Str. 46, 65189 Wiesbaden, Germany

Special Thanks

for the support, help and directional feedback on this work go to:

Technical University of Braunschweig
Institute for Building and Solar Technology
Univ. Prof. Dr.-Ing. M. Norbert Fisch
Dipl.-Ing. Architect Philipp Knöfler

Light design practice Studio DL

And my family, especially my sister Raquel and my brother Earl,
who have been especially supportive.

Contents

List of Figures

List of Tables

1 Introductory Remarks

The history of light goes back a long way. Initially, light was only available in the form of fire, which was an uncontrollable and dangerous element at night. Burning light became safer over thousands of years through the development of candles, oil lamps and gas lamps, but the luminous efficacy was very low. As a result, the large-scale illumination of interior and exterior spaces was associated with a considerable cost factor, which is why for a long time only particularly important buildings and streets were illuminated (cf. Descottes and Ramos 2011).

With the introduction of the electric carbon arc lamp in the middle of the nineteenth century, this changed abruptly.

> The light that flooded a large area was so strong that the ladies opened their umbrellas - not out of respect for the inventors, but to protect themselves against the rays of this mysterious new sun. (La Lumière électrique, 1883, quoted from: Schivelbusch 2004, P. 59, translated)

With this innovative light source, a new, radical way of thinking about light at night began.

The new technology was perceived as a liberation from darkness. Ideas were born such as the idea of illuminating entire cities with a single "artificial sun" (Schivelbusch, P. 11). The new enthusiasm for artificial light did not even stop at areas that are still considered unusual to this day. In fashion, for example, there was the idea of being able to produce glowing jewels (ibid. P. 75).

As a result of this enthusiasm, cities gradually started equipping their streets with electric lights.

At the end of the nineteenth century, after the US American inventor and entrepreneur Thomas Alva Edison had perfected the "light bulb", electric light and the "passion" associated with it moved into private households throughout the country. The "nights" in the cities became brighter and the artificial light became a symbol of progress and prosperity in our culture.

E. E. Goronczy, *Light Pollution in Metropolises*,
https://doi.org/10.1007/978-3-658-29723-7_1

Even if particularly radical ideas were not successful in asserting themselves in every case, there are still many places today which have become a trademark for specific cities precisely because of excessive lighting, and have thus attained cultural importance. Two impressive examples are Piccadilly Circus (London) and Times Square (New York City).

The passion for artificial lighting has meant that we are currently in a situation where cities are literally flooded with artificial light. The general global brightness level at night has increased considerably (Cinzano et al. 2001). The border between day and night is increasingly disappearing in the big metropolises. First scientific experiments show that the excessive luminance levels after dark has negative consequences. This is where terms such as "light pollution" and "light smog" come into play.

Some cities are already trying to counteract this problem with constructional or legal measures. The focus is often on functional street lights (see Marcellino 2014; City of Boston 2015a, b; The City of Toronto 2015).

Globally, however, the majority of people are unaware of the issue of light pollution, making it probably one of the most underestimated environmental pollutants.

This book aims to clarify this issue by dealing with the impact and development of solutions. The book is divided into four main sections.

The first section deals extensively with scientific texts on light pollution and its effects on humans, birds and insects. One key problem that has become apparent through light pollution is the unfavourable effect on the circadian rhythm of different living beings. Hormonal imbalance has been registered, which leads to a series of consequences. The extent of the negative consequences depends mainly on exposure times, and the light spectra and light distribution curves of the luminaires.

The second section includes a comparative analysis of the cities New York City, Boston, Toronto, Warsaw and Hanover. Different urban lighting constellations are explained that have a positive or negative effect on light pollution. Impressions on site and luminance measurements are included in the analysis. With regard to the perception of brightness in urban areas, discrepancies can often be found between the perception of brightness on site and the actual measured values. As a result of the analysis, the initial main problems of the respective cities can be identified. One problem across all the cities is the overexposure caused by clutter situations. Clutter is a term that refers to groups of luminaires which are installed or mounted too close to one another, for example, and as an overall consequence emit unnecessarily high luminance levels.

It is also interesting to see to what extent and in what way the individual metropolises generate light. They differ in many fundamental points, such as the horizontal and vertical dimensions of the cities, the number of inhabitants, the cultural background or the density of buildings.

Moreover, some cities are already implementing measures to reduce light pollution. This is likely to result in a wide range of examples of both good and bad lighting.

The third section deals with the testing of full cutoff luminaires. These are designed to produce little light spill. The tests are carried out using computer-aided lighting simulations. The third section also describes the method developed for evaluating light emissions in urban space. This entails a model-like lighting simulation. The percentage of emissions towards the sky was calculated for the cities under analysis.

Conclusions are drawn from the results of the three main sections, which in turn give rise to the concept solutions defined in the fourth section. The findings contribute to the development of a flow chart, from which solutions for minimizing light pollution can be derived.

The flow chart comprises four blocks: Analysis, Initial measures, Secondary measures and Further measures.

In order to be able to develop particularly efficient solutions, the Analysis is the most important part of the flow chart.

Initial measures are minimal changes to the existing luminaires in the urban space. The measures are therefore low cost, but effective, to achieve a fundamental reduction in light pollution.

The Secondary and Further measures entail structural actions. A lighting master plan is developed to enhance the quality of living in the urban environment and the architectural aesthetic of night-time urban spaces. Light pollution is minimized even further than the effects achieved through the Initial measures.

The lighting master plan mainly concerns street luminaires in urban areas. The actions listed under Further measures also address improving conditions in semi-public buildings.

On the basis of a part of the City of Warsaw, the flow chart is simulated with all measures planned in. It shows that a reduction in light pollution does not necessarily result in a dark urban space. The reduction of light pollution can also be achieved by increasing the quality of stay and enhancing the aesthetic quality of the architecture.

Further summarized results from the work can also be found in issue 04–06/16 of the architecture magazine XIA, intelligente Architektur—Zeitschrift für Architektur und Technik.

Bibliography

Cinzano, P./F. Falchi/C. D. Elvidge, "The first World Atlas of the artificial night sky brightness", in: *Monthly Notices of the Royal Astronomical Society*. Vol. 328, H. 3 (2001), pp. 689–707.

City of Boston, LED Street Lighting, 31.07.2015a, (accessed 10.01.2017). http://www.cityofboston.gov/publicworks/lighting/led.asp

City of Boston, Lights out Boston, 31.07.2015b, (accessed 10.01.2017). http://www.cityofboston.gov/eeos/conservation/lightsoutboston.asp

Descottes, Hervé/Cecilia E. Ramos, *Architectural lighting. Designing with light and space.* New York 2011.

Marcellino, C., Governor signs Marcellino Bill to curtail Light Pollution from state Buildings, 18.12.2014, (accessed 10. 01.2017). https://www.nysenate.gov/newsroom/in-the-news/carl-l-marcellino/governor-signs-marcellino-bill-curtail-light-pollution-state

Schivelbusch, Wolfgang, *Lichtblicke. Zur Geschichte der künstlichen Helligkeit im 19. Jahrhundert*, Frankfurt am Main 2004.

The City of Toronto, *Chapter 629, Property Standards.* 09.07.2015, (accessed 10. 01.2017). https://www.toronto.ca/legdocs/municode/1184_629.pdf

What Is Light Pollution? 2

In general, the term "light pollution" or "light smog" refers to the excessive brightening of the night sky by artificial light (see Dark Sky). Street luminaires, billboards, shop windows and similar lighting installations that are switched on overnight produce light that reaches the atmosphere directly or indirectly (through surface reflections). On the way there, various molecular components of the air are made visible by reflections. The night sky is artificially brightened (Fig. 2.1, 2015).

The intensity of whitening depends on the aerosol composition and can vary on different days or even within a few hours. The resulting "veil of light" over the cities is refered to as a "light dome" or "urban sky glow" (cf. Mizon 2012, pp. 40–47).

The renowned international associations that work against light pollution, "Dark Skies Awareness" and "The International Dark-Sky Association", have added three more key terms to the term urban sky glow: "Light Trespass", "Glare" and "Clutter" (see Dark Skies Awareness; International Dark-Sky Association). Light Trespass refers to incorrectly controlled or incorrectly directed light—i.e. light that radiates into areas where no light is intended and therefore fulfils no purpose. In this context, the term glare means the disturbing glare caused by light sources. Clutter is the expression for groups of luminaires which are positioned too close together, for example, and emit an unnecessarily excessive amount of light as an overall consequence (Fig. 2.2, 2015).

Light in the context of light pollution includes not only the visible range of electromagnetic radiation between 380 and 780 nm, but also the adjacent ranges, infrared (IV) and ultraviolet (UV). IV and UV radiation are located in the visual range of different flora and fauna organisms and thus also have an influence on the environment (Fig. 2.3) (cf. Bartenbach and Witting 2009, pp. 5–11).

In summary, the term light pollution comprises:

E. E. Goronczy, *Light Pollution in Metropolises*,
https://doi.org/10.1007/978-3-658-29723-7_2

Fig. 2.1 Skyline Shanghai 2014. (Source: Philipp Knöfler, unpublished)

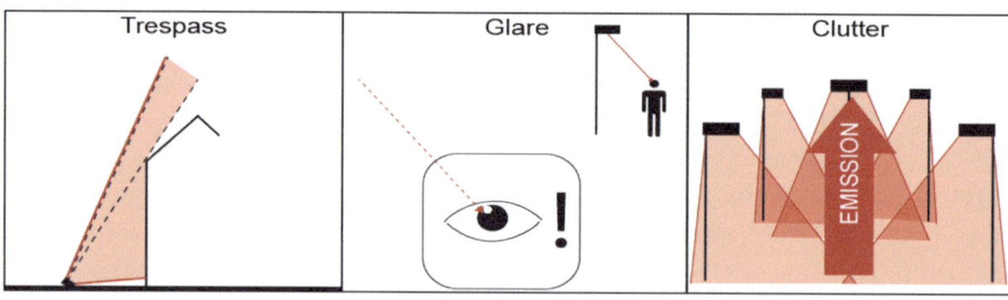

Fig. 2.2 Pictographic explanation of the terms Trespass, Glare, Clutter

- brightening of the night sky
- excess light, both in intensity and spatial terms (light that is unintentionally present in certain areas)
- glare

Light pollution can have far-reaching negative consequences for individual protagonists of the ecosystem. The following chapter illustrates the effect of light pollution on the main victims—humans, birds and insects.

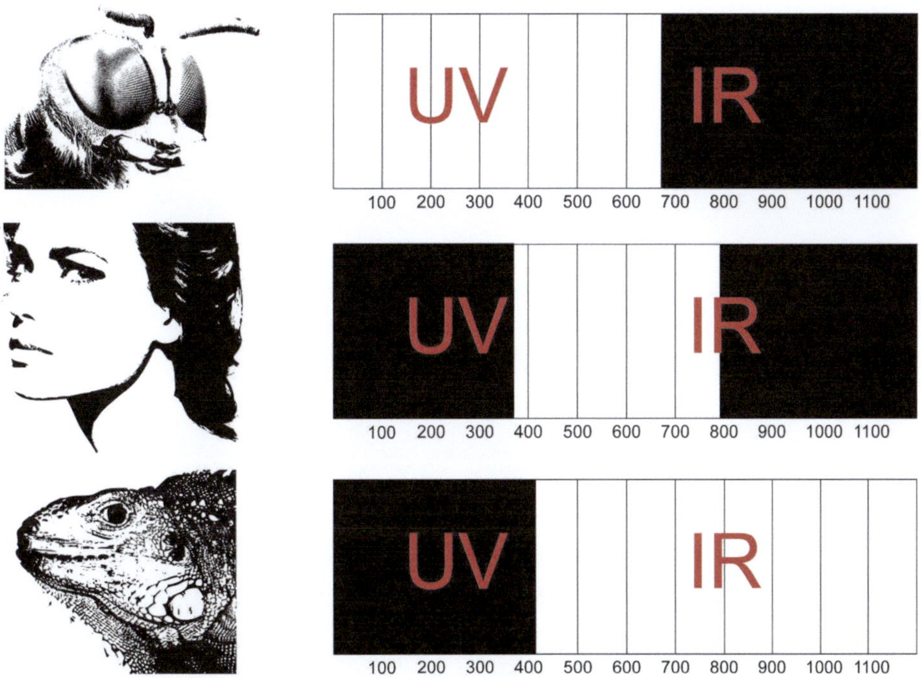

Fig. 2.3 Light perception of insects, humans, reptiles. (Source: Bartenbach and Witting 2009, p. 6, revised by the author)

Bibliography

Bartenbach, Christian/Walter Witting, *Handbuch für Lichtgestaltung. Lichttechnische und wahrnehmungspsychologische Grundlagen*, Vienna 2009.

Dark Skies Awareness, Dark Skies Awareness: Light pollution – what is it and why is it important to know? 21.06.2015, (accessed: 10. 01.2017) http://www.darkskiesawareness.org/faq-what-is-lp.php

Fachgruppe Dark Sky, Dark Sky – Initiative gegen Lichtverschmutzung, 21.06.2015, (accessed: 10. 01.2017) http://www.lichtverschmutzung.de

International Dark-Sky Association, Light Pollution Matters, 22.06.2015, (accessed 10.1.2017) http://www.darksky.org/light-pollution-topics/light-pollution-matters

Mizon, Bob, *Light pollution. Responses and Remedies* second edition, New York, NY 2012.

3 Influence and Effects of Artificial Light After Dark and Light Pollution

3.1 ... on the Human Being

Cinzano et al. (2001) illustrate the intensity with which artificial light is applied in and around cities after dark. In the study carried out between 1996 and 1997, twenty-eight photographs of the earth were taken and assembled into a map.[1] Interfering factors such as overexposure, fires or unclear atmospheres were removed from the images so that only images of clear skies were processed in the end result (Fig. 3.1).

The resulting map shows the relationship between artificial and natural sky brightness. The degree of artificial lighting is indicated via false colours.

<11%	Black
11–33%	Blue
1–3%	Yellow
3–9%	Orange
>9%	Red

This shows that North America, Europe and Japan in particular are strongly affected by artificial night sky brightness. Cinzano and his colleagues go one step further in their analysis. Comparing their map material with the United States Department of Energy (DOE) population density data base they were able to determine the fraction of population

[1]The pictures were taken with the aid of a satellite and a special camera. The spectral capacity of the camera was between 440 and 940 nm, with particularly high sensitivity in the 500–650 nm range. This was of particular benefit to the photographs, given that the street lights commonly used at that time had a spectrum of between 545 nm and 630 nm (mercury vapour lamp: 545 nm and 575 nm; high-pressure sodium vapour lamp: 540–630 nm; low-pressure sodium vapour lamp: 589 nm) (cf. Cinzano et al. 2001, p. 690).

E. E. Goronczy, *Light Pollution in Metropolises*,
https://doi.org/10.1007/978-3-658-29723-7_3

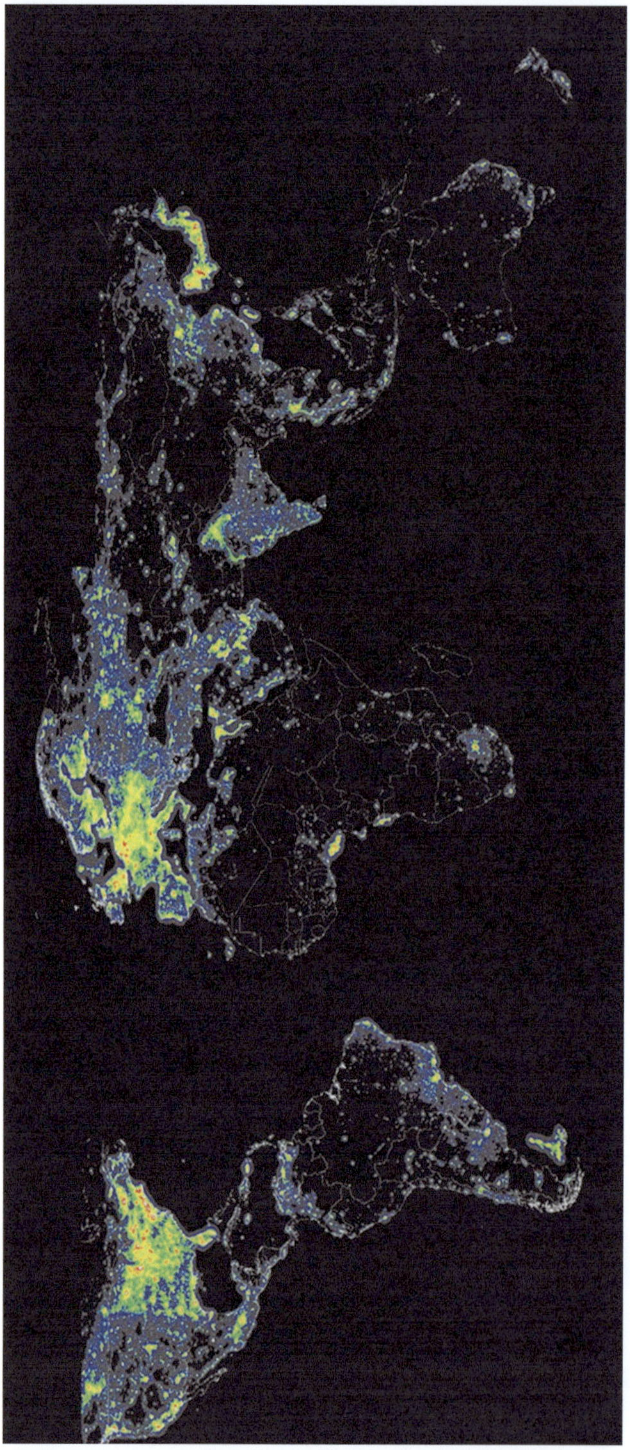

Fig. 3.1 The first World Atlas of the artificial night sky brightness. (Cinzano et al. 2001, p. 690)

exposed to a sky of a given brightness in any region. Cinzano et al. have generated data for over 180 countries. An excerpt of the resulting evaluation is shown in Table 3.1.

In Germany, 94% of the population therefore live under a night sky, which is 100% brighter than the natural night sky (natural night sky: approx. 200 μcd/m^2). 25% of the German population even live in areas with a sky brightness that is nine times higher than the natural brightness of the night sky.

100% of the population in Germany, Italy, Great Britain and Japan, and 99% in the USA and the European Union, live in areas which are brighter than the threshold value for natural sky brightness at night. The average brightness of the night is 68% above 890 μcd/m^2 in the European Union and 81% in the USA, which corresponds to the brightness of the night at full moon. These areas therefore never experience a "dark" night.

The values relating to the night adaptation of the eyes should also be emphasized. As a result, 44% of the population in the USA, 41% in Japan, and even 59% in Canada are exposed to a level of brightness that hinders or prevents night vision.

Table 3.1 Percentage of the population living under a certain night sky brightness

Country	$\geq 0{,}11b_n$	$\geq 0{,}33b_n$	$\geq b_n$	$\geq 3b_n$	$\geq 9b_n$	$\geq 27b_n$	$\geq b_{FM}$	$\geq b_{mw}$	$\geq b_e$
Australia	71	69	68	62	37	1	60	48	8
Canada	97	94	90	83	71	46	82	77	59
China	54	41	29	13	5	1	12	7	2
Germany	100	100	94	66	25	0	60	40	5
Italy	100	99	95	78	35	6	72	50	15
Japan	100	99	96	86	63	27	84	73	41
Poland	99	88	72	44	18	0	39	26	3
Russia	87	80	73	60	34	8	57	44	15
Spain	98	93	87	78	57	25	76	67	38
UK	100	98	94	79	40	4	74	55	15
USA	99	97	93	83	62	30	81	71	44
European Union	**99**	**97**	**90**	**72**	**38**	**8**	**68**	**51**	**17**
The world	**62**	**53**	**43**	**30**	**16**	**6**	**28**	**21**	**9**

$\geq 0.11b_n$ The brightness of the sky is at least 11% higher than that of the natural sky
$\geq 0.33b_n$ The brightness of the sky is at least 33% higher than that of the natural sky
$\geq b_n$ The brightness of the sky is at least 100% higher than that of the natural sky
$\geq 3b_n$ The brightness of the sky is at least 3 times greater than that of the natural sky
$\geq 9b_n$ The brightness of the sky is at least 9 times greater than that of the natural sky
$\geq 27b_n$ The brightness of the sky is at least 27 times greater than that of the natural sky
$\geq b_{FM}$ Sky brightness is brighter than the natural sky with a full moon ($\geq$890 μcd/m^2)
$\geq b_{mw}$ It is no longer possible to see the Milky Way ($\geq 6b_n$)
$\geq b_e$ The eye no longer has the ability to adapt to night vision under the night sky ($\geq$ 4452μcd/m^2)
Source: Cinzano et al. 2001, pp. 698–700, selected values

The study by Cinzano et al. is now over 15 years old. The calculated growth rate of light pollution at that time was 7–10% per year for the European Union alone (cf. Cinzano 2002, p. 91–101). Thus the values are much higher at the present time. Moreover, countries such as China or the United Arab Emirates, who experienced their "economic boom" at a later date, could not be taken into account with regard to the growth rate. As a result, the growth rate will be higher than previously forecast.

3.1.1 Immediate and Long-Term Impairments

Artificial light at night has multiple effects on the human organism. Some are directly perceptible, others take place at the molecular level and are therefore part of a slower process.

Glare

One of the immediately noticeable effects is excessive glare from light sources. DIN EN 12665 describes the term glare as follows: "unpleasant visual condition due to unfavourable luminance distribution or excessive contrasts" (Deutsches Institut für Normung 2011, translated). In the international context, the two terms "Discomfort Glare" and "Disability Glare" make it easier and more precise to illustrate the subject of glare.

The human eye has the ability to adapt to the luminance level of the environment. This is done with the aid of the pupil, which opens or closes depending on the brightness of the surroundings and thus optimally regulates the incidence of light. Glare occurs when light sources appear in the field of vision whose brightness level is considerably higher than the brightness level of the environment to which the eye has adapted (see Philips Lighting Academy 2008, pp. 33–34). This can be perceived as disturbing. If a deteriorated visual performance cannot be proven by external measurements, one speaks of "psychological glare" or "discomfort glare" (cf. Reidenbach et al. 2008, pp. 74–84).

Psychological glare is a purely subjective sensation in which the visual ability is not diminished (cf. Noyes 2001, pp. 81–82). It is essentially harassment. The consequences of this can be constant changes in viewing direction, sensation of stress, fatigue and discomfort (cf. Reidenbach et al. 2008, p. 83).

The second form of glare results from the consequence of the ability to adapt and the structure of the eye. If one looks directly into a bright light source, the eye adapts to the brightness level (however, it should be noted here that the pupil diameter does not decrease due to the brightest object in the visual field, but due to the total amount of light in the visual field). The reduced diameter of the pupil means that not enough light for the darker surroundings can enter the eye. The perception of the darker areas in the field of vision is therefore only possible to a limited extent (cf. Narisada and Schreuder 2004, pp. 296–307).

In most cases, however, the main cause of glare when scattered light enters the eye. This is caused by the opaque parts of the cornea, the lens and the vitreous body, which prevent visual perception. Older people have a lower glare threshold because the advanced opacity of the lens increases the amount of scattered light that enters the eye (cf. Hentschel 1994, pp. 63–65).

This is a common phenomenon and often occurs at night, for example with oncoming headlights in road traffic, and is known as "physiological glare" or "disability glare" (cf. Schlag et al. 2009, pp. 32–33).

While psychological glare can be a nuisance, physiological glare has a measurable effect on the eye and vision. The effects of this glare can lead to a reduction in the sensitivity to differences in visual acuity, form and design ability, depth perception and perception speed (cf. Institute for Occupational Safety and Health of the German Accident Insurance 2010, pp. 3–5).

Since the development and widespread use of LEDs, especially in recent years, more and more physiological forms of glare have appeared. The LED has a significant peak in the blue part of the spectrum (Figs. 3.4 and 3.9). The eyes react sensitively to this area, especially at night. As a consequence LED light is perceived as particularly dazzling (cf. Bullough et al. 2003, pp. 21–25). For example, due to the comparatively higher blue component, xenon headlamps also subjectively appear to be brighter than halogen headlamps with the same output.

A further effect resulting from the high blue component is that it increases the time required for the eyes to adapt to the dark: the shorter the wavelength, the longer the adaptation time. Accordingly, the eyes adapt faster to red light (long-wave light) than to blue light (short-wave light) (cf. Bartlett and Graham 1965).

Another effect of the blue component of the LED is described by Barbur et al. According to their research, light with a blue component reduces the size of the pupil. A decrease in pupil dilation in combination with a low brightness results in vision impairment, as too little light can be detected by the eye to perceive the surroundings (cf. Barbur et al. 1992, pp. 137–141).

These immediately noticeable impairments correspond to avoidance behaviour, because they also occur under natural conditions. For example, in the case of strong glare we turn our heads away from the light source, like a reflex. That said, the stress factor caused by artificial light sources should definitely be avoided.

Influence on Circadian Rhythm

The situation is different when it comes to the successive, barely noticeable consequences of light pollution. The effects take place on a hormonal level within the human organism and are the result of night-time lighting.

Most organisms have an "inner clock", referred to as the circadian rhythm. This rhythm regulates various processes, for example the production of different hormones in humans at specific times of the day (cf. Salvendy 2012, p. 694) or the regulation of photosynthesis in plants (cf. Esser 2007, pp. 299–303). The inner clocks that determine the circadian rhythm are mainly the different brightness levels over the course of the day and night phases, and the different wavelengths of light.

A full moon can generate an illuminance of approximately one lux (see Yoshizawa 2015, p. 13). According to Wright et al. an illuminance of 1.5 lux can already impact the circadian rhythm (Wright et al. 2001, pp. 14027–14,032). In regions where night sky brightness is brighter than at full moon ($\geq b_{FM}$), there is a danger that the circadian rhythm will suffer interference. According to Cinzano et al. (2001, see Table 3.1) this would apply to more than half of the population in Germany and Europe, and even more than 80% in Japan, Canada and the USA.

This data proves that artificial light at night interferes with the circadian system of a large number of people, which can have serious health consequences.

In order to be able to deal more precisely with the effects of artificial light at night, it is necessary to explain the processes in the eye and in the body in more detail.

Circadian Rhythm and Hormone Production

The eye has three different light receptors on the retina: the cones, the rods and the ganglion cells. While the cones are responsible for day vision and colour perception and the rods are responsible for the task of night vision, the ganglion cells only perceive light stimuli (light and dark) and do not serve to perceive forms and colours. The ganglion cells are distributed over the entire retina and have a special sensitivity in the lower part of the eye (Fig. 3.2). Depending on the brightness of the environment, they regulate biological processes in the body such as the pupil reflex or the secretion or suppression of various hormones (see Fördergemeinschaft Gutes Licht 2014, pp. 14–15). One of the most important hormones in this context is melatonin.

Melatonin is produced in the pineal gland in the evening at low levels of brightness. The hormone initiates the resting phase of the body and is responsible for decelerating many metabolic processes. It also causes fatigue. When the brightness level increases in the morning, the ganglion cells are stimulated by light, which reduces melatonin production.

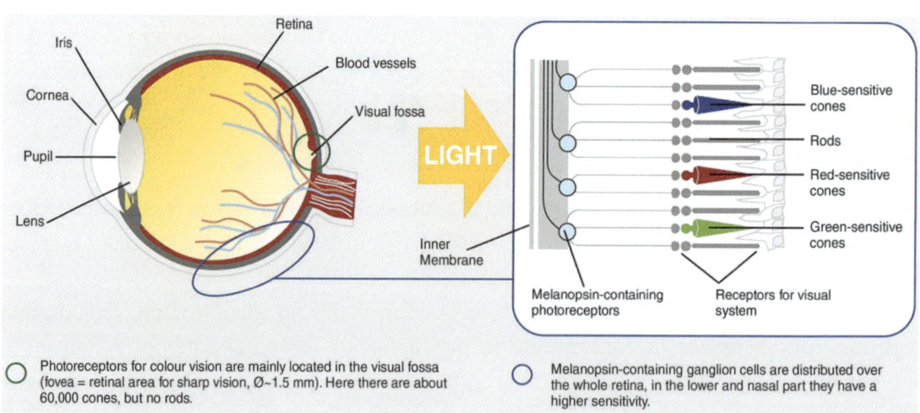

Fig. 3.2 Sensitive ganglion cells. (Source: licht.de, sense organ eye; access: 10.01.2017. https://www.licht.de/de/trends-wissen/ueber-licht/licht-und-sehen/sinnesorgan-auge/)

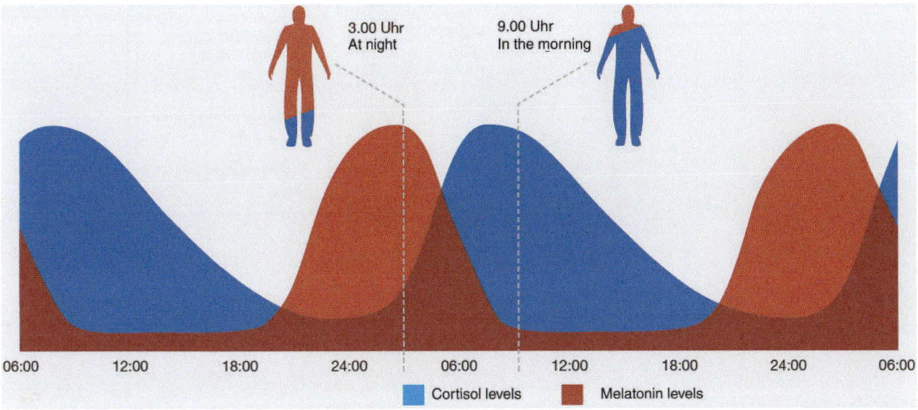

Fig. 3.3 The influence of daylight on hormone levels. (Source: licht.de, light clocks the internal clock; access: 10.01.2017. https://www.licht.de/de/trends-wissen/licht-specials/human-centric-lighting/mensch-und-licht/licht-taktet-die-innere-uhr/)

At the same time, an increased secretion of the hormones serotonin and cortisol occurs. These hormones have a mood-enhancing and motivating effect. The body and its functions are reactivated by serotonin. If the brightness level decreases again in the evening, melatonin production starts again, whereupon the serotonin level decreases with the cortisol level (Fig. 3.3). This roughly describes the circadian cycle.

An experiment carried out in 2005 showed that the ganglion cells react differently to wavelengths (cf. Cajochen et al. 2005, pp. 1311–1316). In the experiment, the participants were exposed to monochromatic light at different wavelengths for 2 h at night. Melatonin levels were compared before and after irradiation. It was found that subjects exposed to short-wave radiation produced less melatonin than subjects exposed to long-wave light. Significant melatonin suppression took place particularly in the bluish range at 460 nm, whereas wavelengths >550 nm had no influence on melatonin production. The circadian rhythm is therefore not only controlled by brightness, but also by the wavelength of the rays.

Interesting in this context is the spectral composition of the cool white LED. This has an increased emission in the sensitive wavelength range at 460 nm, with which this light source undoubtedly influences the circadian rhythm (Fig. 3.4). This type of LED is increasingly used in street lighting today. LEDs with this light composition are particularly energy-efficient, which makes them very attractive to cost-conscious planners and they are therefore frequently specified (cf. Chen 2010, pp. 8–12).

Intervention in the cycle of melatonin levels, through light after dark for example, can have many different effects. Among other things, melatonin is a cancer blocker and strengthens the immune system. If too little melatonin is produced and released during the day, or at night, this function is reduced accordingly.

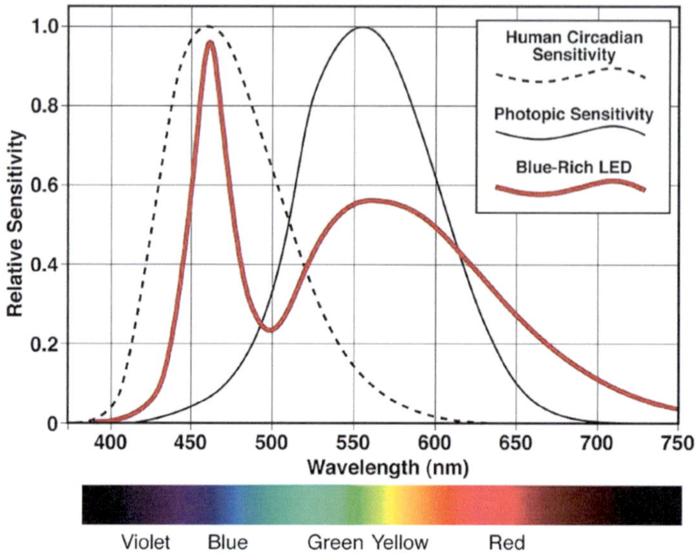

Fig. 3.4 The circadian sensitivity curve and the photopic perception curve (night vision) in relation to the spectral range of the cool white (blue-rich) LED. (Source: International Dark-Sky Association 2010, p. 12, edited by the author)

Especially the risk of breast cancer in women is suspected to increase if the melatonin level is artificially suppressed (cf. Bullough et al. 2006, pp. 375–383; Glickmann et al. 2002, pp. 17–22; Kloog et al. 2008, pp. 65–81; Stevens et al. 2007, pp. 1357–1362).

A case-control study carried out in Denmark further confirms this suspicion. A total of 7035 women took part in the study. The volunteers were both healthy women and women with breast cancer who worked mainly at night. The women were aged between 30 and 54 years. An individual work history was reconstructed for each participant going back as far as 1964. This clearly revealed tendencies indicating that the probability of developing breast cancer increases the longer night work was carried out (cf. Hansen 2001, pp. 74–77). In conclusion, the longer one suppresses the natural circadian rhythm and melatonin level, the higher the risk of developing breast cancer. An essential factor that leads to melatonin suppression is artificial lighting after dark and how this undermines the dark night sky. The luminous intensity of a gently glowing light source at night reduces the melatonin level by 50% after only 39 min (cf. Schulmeister et al. 2004).

How dangerous melatonin suppression actually is cannot be precisely defined at the present time. Many factors play a role in cancer, and melatonin suppression associated with light pollution is only one of them.

What is certain, however, is that melatonin has a decisive influence on cancer growth and development. Cancer tumors were implanted in rats during an experiment performed by the National Cancer Institute in Maryland. Some of the rats were treated with melatonin. The cancer growth of these rats was significantly slower than that of the untreated animals (cf. Tamarkin et al. 1981, pp. 4432–4436).

Other experiments with rats and chickens have shown that melatonin suppression at night also has an effect on metabolism and the development of diabetes, overweight and heart attack (cf. Robbins et al. 1984, pp. 269–277; Rodríguez et al. 2007, pp. 15–24; Mustonen et al. 2002, pp. 716–723). First studies prove a plausibility of these connections also with humans (see Haus und Smolensky 2006, pp. 489–500; Brugger et al. 1995, p. 1408; Bass and Turek 2005, pp. 15–16).

With regard to light pollution, the suppression of natural melatonin levels is the most serious danger. But the manipulation of the circadian rhythm also has effects on other substances and hormones such as:

- Prolactin (cf. Vaticon et al. 1980, pp. 277–288), growth hormone for the mammary gland (important in pregnant women).
- Glucocorticoids (cf. Leproult et al. 2001, p. 151–157), steroid hormone (influences metabolism, water and electrolyte balance, cardiovascular system and nervous system; also has anti-inflammatory and immunosuppressive effects).
- Adrenocorticotropic hormones (cf. Fischmann et al. 1988, pp. 309–316) are responsible for the release of cortisol and corticosterone.
- Corticotropin-releasing hormone (ibid.), stress-inducing hormone (but is also involved in the cardiovascular system, inflammatory processes, reproductive system, pregnancy, thermoregulation, food intake and psyche).

However, little research has yet been done into the consequences of this influence.

To summarize, however, it can be said that some of the consequences of artificial lighting at night are avoidable. For example, it is possible to adapt one's daily habits to align with the circadian rhythm and to "block out" spill light from the streets after dark by taking one's own measures (e.g. by hanging light-excluding curtains at the windows of one's home).

The main points of this section are summarised in Table 3.2.

Table 3.2 Brief summary of the effects of light pollution on humans

Cause	Impact
glare caused by the light source	• uncomfortable, harassing effect • visual impairment
LED spectrum (high blue component)	• looks unnaturally bright • the eye takes longer to adapt • suppresses melatonin production
brightening of the night sky (sky glow)	• suppresses secretion of different hormones in metabolic system, including melatonin; consequently suspected of promoting: – diabetes – overweight – cancer – myocardial infarction (heart attack)

3.2 ... on Birds

In contrast to humans, animals do not have the possibility to escape light pollution. They are freely exposed to it and in the case of birds, artificial light is usually even attractive.

Numerous records document mass deaths of birds caused by light pollution. A spectacular case is described by Johnston and Haines. According to the study, 75,000 migratory birds died in a single night in 1954. This was caused by floodlights at two nearby airfields. The floodlighting influenced the orientation of the birds, which gave rise to mass collisions and birds becoming disoriented and dying from exhaustion (cf. Johnston and Haines 1957, pp. 447–458).

There are similar cases in Germany. For example, in the period from October 2006 to November 2007 about 1000 dead birds were observed close to the "Post Tower" in Bonn. The cause of the mass death was the facade lighting of the high-rise building (cf. Haupt 2008).

The extent of the negative impairment of artificial light at night is so immense that associations such as FLAP (The Fatal Light Awareness Program, Canada) or NABU (Naturschutzbund Deutschland) are already advocating bird-compatible lighting and measures.

Light at night has a high importance for birds. Up to 60% of migratory birds, for example, migrate at night (cf. Schmaljohann et al. 2007). The light essentially has two different functions. On the one hand, luminous celestial bodies such as the moon and the stars serve as spatial orientation, and on the other hand light has a temporal, periodical function (cf. Böttcher 2001, pp. 19–20.)

3.2.1 Orientation Ability

Migratory birds orient themselves at night either by the magnetic field of the earth or, as already explained, by the positions of the stars and the moon (cf. Berthold 1991). Mainly the visual orientation (stars and moon) is used.

The artificial brightening of the night sky, however, renders visual orientation barely possible, especially in urban areas. The contrast between the starry sky and the city (as a result of sky glow) is too great. The stars cannot be perceived. This is verified in the investigations carried out by Cinzano et al. (2001) (see Sect. 3.1). If visual orientation is not possible, the birds still have the possibility to orient themselves by the magnetic field of the earth.

Strong artificial light sources are another disturbing factor for orientation. Birds are strongly attracted by it, in particular under bad visibility conditions. Glare is an especially dangerous issue. Birds often react by flying directly towards the bright light source and colliding with the luminaire housing or the highly illuminated structure. At speeds of 75–120 km/h, the collision is often fatal (see Creutz 1989).

Fig. 3.5 Ground zero memorial, skybeamer. (Source: David Ringer/Audubon. http://www.dailymail.co.uk/news/article-3232091/Tribute-Light-shut-four-times-THOUSANDS-migrating-birds-trapped-like-moths-powerful-beams-produced-Earth-commemorate-victims-9-11.html)

Differences between pulsating (blinking) light sources and light sources that emit constant light have also been detected. For example, an experiment in North America has shown that flashing lights have less influence on migratory birds than constant lights (cf. Evans et al. 2007, pp. 476–488).

In general, weather conditions are also a decisive factor: the cloudier the sky, the stronger the rays of light and the brighter the sky. Consequently, the effects are also more drastic (cf. Poot et al. 2008).

If birds then resort to the magnetic field as a means of navigation to overcome all these impediments to their visual orientation, manifold impediments lurk here as well. The colour of the light plays an important role here. Green light has the least influence, whereas red, white and yellow light can lead to strong disorientation (cf. Poot et al. 2008). Yellow light in particular causes a lot of confusion. Birds tend to orbit the light, often resulting in fatal mass collisions (Fig. 3.5). It is assumed that this behaviour occurs because the short-wave blue and green parts of the spectrum are only present to a small extent given a yellowish light source. But this blue and green part of the spectrum can be usefull for magnetic orientation. Its absence can lead to a lack of orientation (see Rich and Longcore 2006). Accordingly, particularly in areas with massive sky glow, i.e. primarily large cities, this can lead to serious consequences—birds dying of exhaustion having been disoriented for long periods of time.

3.2.2 Influence on the Circadian Rhythm

Another serious problem is the intervention in the circadian rhythm of birds, similar to that in humans. Artificial lighting interferes with phenological activities. This leads to changes in birds' physical condition, reproductive activities and hunting behaviour.

An important instrument of songbirds is their singing in the morning and evening. Among other things, it serves to attract partners and to mark their territory. Singing also enables the female to judge the male's physical condition, health and performance. Singing is therefore one of the most important instruments in the choice of partner and reproduction (cf. Birkhead and van Grouw 2015, pp. 27–58).

The artificial light at night prolongs the day. Urban songbirds singing in the morning start singing earlier than songbirds of the same species in forest areas. Davide Dominoni and his team of international scientists and researchers were able to determine a difference of 17–29 min (cf. Dominoni et al. 2013b, pp. 1–7). The exact effects of this behaviour are still relatively unknown at the present time. On the one hand, observations made by the Max Planck Institute (Kempenaers et al. 2010) suggest that by prolonging the day and singing early, the chance of mating and the time frame to go looking for food are increased (cf. Kempenaers et al. 2010, pp. 1735–1739); on the other hand there is speculation that the danger of becoming prey also increases accordingly. In addition, prolonged daily activities can have a negative effect on the birds' physical condition (physical exhaustion) and thus also on the quality of the singing. As a result, the search for a reproductive partner becomes more difficult due to a lack of vocal attractiveness (cf. Da Silva et al. 2015), i.e. weakened birdsong.

During a two-year experiment at the University of Glasgow (Dominoni et al. 2013c), the effects of night-time brightness were investigated more closely under laboratory conditions. For this purpose birds were divided into two groups. One group (group A) was exposed to a brightness of 0.3 lux every night, which corresponds to the degree of artificial lighting in large cities at night. The second group (group B) was not exposed to light. Blood was taken regularly from both groups in order to determine hormone levels.

In the first year it was recorded that group A was more active than group B. Daily activities such as courting songs or searching for food began accordingly earlier and stopped later. Sexual responsiveness also started prematurely compared to group B. Thus, the testicles and certain sex hormones developed earlier than usual for the time of year. This actually had a positive effect on reproduction (cf. Dominoni et al. 2013c, pp. 1–9). The same observations were made as in the report compiled by the Max Planck Institute. Up to this point the developments could therefore be interpreted as an advantage vis-à-vis other birds.

In the second year, however, it was discovered that the daily exposure of group A had long-term consequences. In the second half of the experiment, reproduction stagnated at first, until reproduction ceased altogether. There is no concrete explanation for this development to date (ibid.). A state of stress that occurring due to the daily overexposure to artificial light could be a logical explanation.

Whether the longer daily cycle is advantageous or disadvantageous is difficult to judge in conclusion. The observations made by the Max Planck Institute and the University of Glasgow should only be compared to a limited extent. All data from the Max Planck Institute was collected outdoors in the natural habitat of birds. The experiment at the University of Glasgow, on the other hand, was a laboratory experiment. The unnatural environment for the birds will have been included in the experiment and, for example, will have increased the stress factor. In addition, Group A was deprived of the opportunity to avoid exposure, which is partly possible in the wild due to migration. A decline in reproduction in the second year might not have occurred at all. In order to be able to make clear statements on the subject, further experiments are required.

It should not be denied, however, that exposure to artificial light at night also inhibits melatonin production in birds (Dominoni et al. 2013a, pp. 1–10). The hormone is particularly important for migratory birds. Among other things, it controls the timing of migration and supports orientation. If the melatonin level is impaired, corresponding functions can be disturbed (cf. Gwinner 1996, pp. 47–63; Schneider et al. 1994, pp. 47–63).

Another factor influenced by the unnatural length of the day is the time of hatching. Birds judge in which season they are on the basis the length of day and night. This also determines the time of hatching. The actual weather only plays a subordinate role (cf. Lambrechts et al. 1997, pp. 5153–5155; Dawson et al. 2001, pp. 365–380).

Another experiment revealed that in European cities blackbirds start hatching on average 19 days earlier than their fellow blackbirds in the forest (see Dominoni and Partecke 2015). Another analysis even came to the conclusion that hatching is sometimes one month too early (cf. Partecke et al. 2005, pp. 295–305).

In an experiment carried out by biologist William Rowan, a junco bird was encouraged to breed in the middle of winter (cf. Rowan 1925, pp. 494–495) simply by artificially prolonging the day (by providing light at night). This experiment shows how much influence light has on the timing of activities in the bird world.

Premature hatching can have a negative impact on the offspring, especially during the cold winter months when it is difficult to find food.

The search for food in general is also influenced by artificial lighting after dark. Given the "longer" days, the time spent searching for food also increases. This has an impact on the entire ecosystem. Close observation of short-eared owls showed how predator-prey behaviour adapts to the lighting conditions. Contrary to the majority of birds, owls are nocturnal animals. Bright nights, and the consequently enhanced visibility conditions, have prolonged the time available for hunting. The deer mouse, favourite prey of the short-eared owl, has adapted to this by reducing the time it spends searching for food itself. This minimizes the risk of being captured, but also the time allowed for food intake. It can be

Table 3.3 Brief summary of light pollution effects on birds

Cause	Impact
Light source	• Attracted by luminaires and often fly into them, killing themselves. • The rays of light come across as attractive, resulting in mass collisions. • The rays of light undermine the birds' ability to orient themselves.
Light spectrum (yellowish)	• Can be disturbing for the birds' sense of direction.
Artificial light at night (sky glow)	• Orientation by the stars is inhibited. • Negative impact on circadian rhythm and consequently on: – Hunting behaviour – Breeding times – Courtship behaviour – Physical condition • Melatonin suppression, disrupting the temporal orientation of migratory birds.

assumed that such behavioural changes give rise to further adaptation and changes in other links in the ecosystem, resulting in a chain of ecological changes, the effects of which are difficult to gauge.

The most important facts mentioned in the last section are summarized in Table 3.3.

3.3 . . . on Insects

There are an estimated 30 million different insect species on earth, making them one of the most species-rich groups in the animal kingdom. In addition, they have numerous important functions in the ecosystem, e.g. in the food chain and in flower pollination (cf. Hickmann and Weber 2008, p. 622). It cannot therefore be ruled out that the effects of light pollution on insects in general also brings about changes in the overall ecological context. Insects should therefore be accorded high significance.

Insects react to light pollution in a variety of ways, depending on the species. Due to the diversity of species, it is impossible to make statements about every insect species. Therefore, only the majority reaction of nocturnal insects to light pollution will be discussed in the following section.

3.3.1 "Attraction" Through Different Types of Luminaires

In terms of light pollution, the luminaire itself is the main hazard for flying insects. Artificial lighting at night has a comparatively small effect on the insect world.

The reason why nocturnal flying insects are attracted to light sources at all, usually encircling them in a spiral, is not always clear. However, there are numerous theories. Some of these theories are presented below:

- *March-Band-Theory*
  Flying insects try to escape the light by flying to the point with the greatest light-dark contrast. This point is located directly at the border between light and darkness. As a result, the insect cannot escape the light beam emitted by a lamp and flies in a spiral path around the lamp (see Henke 1991).
- *flight safety theory*
  Insects fly directly to the light source. There can be no obstacles blocking the insects' direct visual contact with the light source. The danger of a collision with an object is thus avoided when flying directly to the light source (see Cleve 1967, pp. 33–53).
- *lunar orientation theory*
  Flying insects orient themselves by the moon. Given its distance to the earth, the moon is seen as a stationary object in the sky. Consequently, when moving around on Earth, the Moon does not appear to change its position. If insects wish to fly in one direction, to support orientation the moon must always be to their left. Light sources such as street lights can be confused with the moon because of their intense luminosity. If an insect orients itself according to the light emitted by a street lamp, it will end up orbiting around the light source, since it must always be on the same side (see Naturwissenschaftlicher Verein Wuppertal 2015).

There are several factors that determine the level of attractiveness of a light source for insects.

In principle, the location of a luminaire already plays an important role. Luminaires in forest areas have a greater attraction potential than luminaires in urban areas. This is because the biodiversity and total number of insects in the forest are considerably higher than in the city. Consequently, more insects can be influenced by the light source (cf. Böttcher 2001, p. 64).

In 1993 Bauer was able to establish that the height of the source of light also had an influence on attracting insects. For example, luminaires at a height of 5 m attract 1.5 times more insects than luminaires at a height of 2.5 m (cf. Bauer 1993).

Since insects orient themselves according to the highest contrasts in space (March-Band-Theory), luminaires in close proximity to each other attract fewer insects than luminaires that are positioned further away from each other. The reason for this is that the overlap of the light cones is reduced with increasing distance between the individual luminaires. This reduces the uniformity of the lighting, which increases the background contrast between the luminaires. Luminaires that are so far apart from each other that "islands of light" are created. They provide a great contrast to the background scenario and are therefore preferred (cf. Frank 1988, pp. 63–93; Kurtze 1974, pp. 297–344).

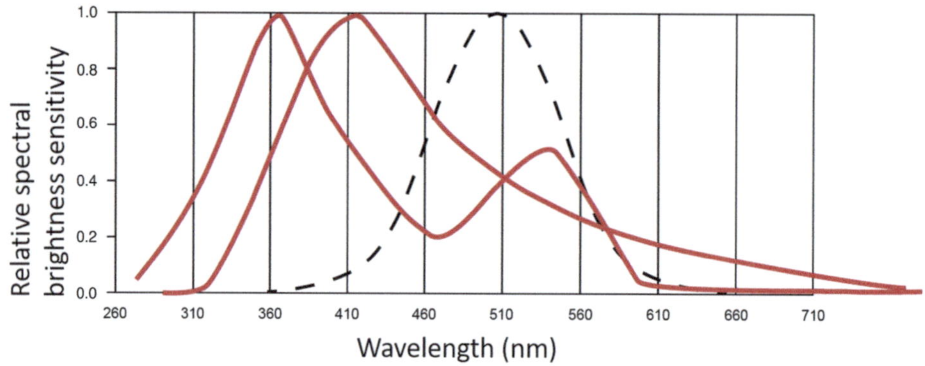

Fig. 3.6 Relative spectral sensitivity of insects (red) and night vision in humans (dotted line). (Source: Bartenbach and Witting 2009, p. 11, edited by the author)

The attraction distance of luminaires is estimated to be 20–700 m, whereby the weather (e.g. wind, rain and temperature) can considerably influence this (cf. Böttcher 2001, pp. 19–52, 75–100). The design of the luminaire heads is also a crucial factor: full cutoff luminaires are approached much less frequently than spherical luminaires (non-cutoff luminaires). Full cutoff luminaires prevent glare from the light source. Scattered light is thereby prevented to a large extent and the light source itself can only be seen when viewed from certain angles. This reduces the long-distance effect and thus the attraction potential (see Bauer 1993).

However, the light source with its specific spectrum plays the most important role in attracting insects. Insects have a visible spectrum that reaches into the UV range (see Fig. 2.3). The short-wave range is particularly attractive for insects at night, since most insects have a maximum perception of brightness of between approximately 350–385 nm and 385–435 nm (UV, magenta, blue parts of the spectrum). The maximum perception of brightness in humans under night vision is still in the long-wave range, at approximately 507 nm (Fig. 3.6) (cf. Bartenbach and Witting 2009, p. 8). Thus it can happen that luminaires that appear less bright to the human eye are extremely bright for insects.

As a result, studies prior to the appearance of LEDs in street lighting have shown that mercury vapour lamps and metal halide lamps in particular are highly attractive to insects. This is a result of the high blue component in the spectrum, part of which lies in the UV range.

High-pressure sodium vapour lamps, on the other hand, have hardly any blue components, which means that they are the least frequented by insects (Fig. 3.7, the blue, violet and UV components lie in the range < 490 nm).

Luminaires featuring short wavelength light have a six to ten times greater chance of attracting insects than those equipped with light sources with longer wavelength light. Give the way they perceive brightness, the majority of insects perceive sodium lamps as being around 90% darker than we humans perceive them, whereas mercury vapour lamps are

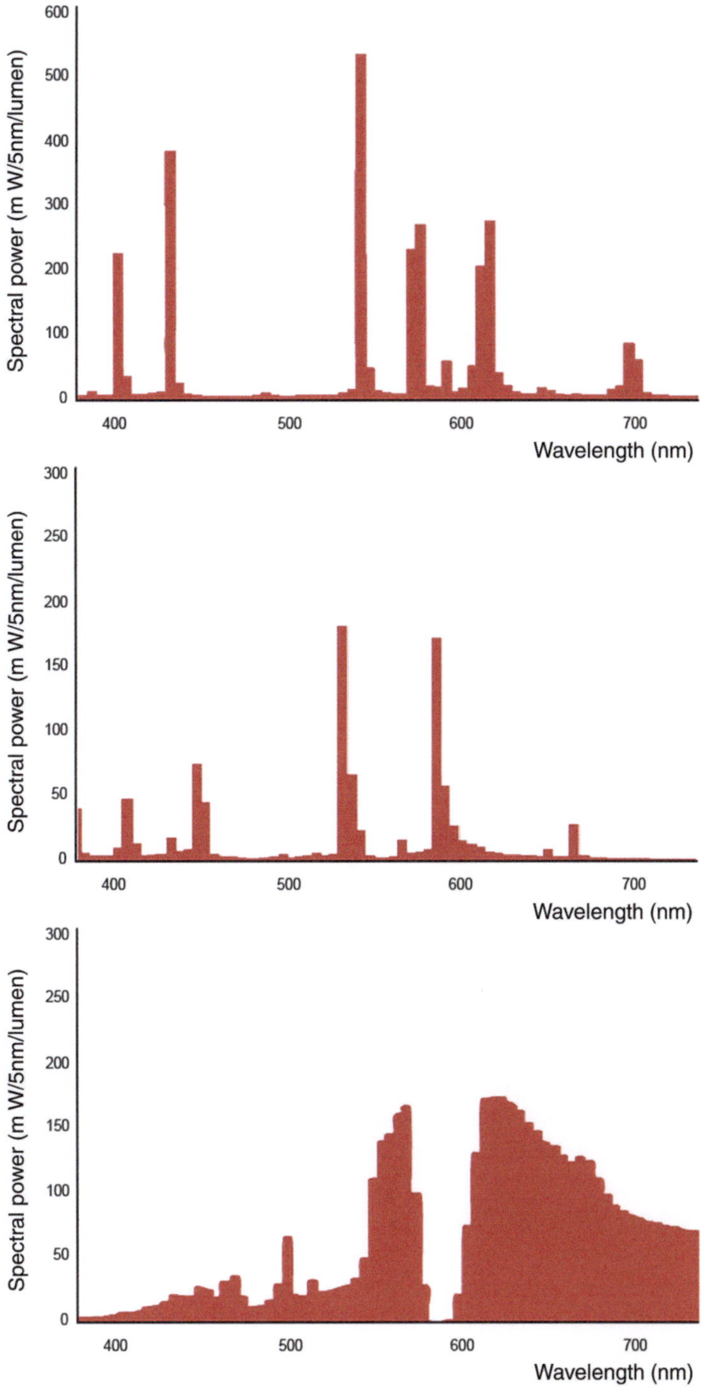

Fig. 3.7 Spectral composition of mercury vapour lamps, metal halide lamps, high-pressure sodium vapour lamps (from top to bottom). (Source: LMK Software)

seen by the majority of insects as being nine time brighter than we humans see them (cf. Steck 1997; Cleve 1967; Autrum et al. 1979, pp. 503–580).

Since the increased use of LED technology in road space, the term "insect-friendly light" has been used more and more frequently, partly by the industry, but also, for example, by the German Association for the Environment and Nature Conservation (Bund für Umwelt und Naturschutz Deutschland 2015). The extent to which LED light is "insect-friendly light" still needs to be clarified.

Reference is especially made to the fact that LEDs have no UV component in the spectrum, so insects that react sensitively to UV are not attracted. Consequently, LED luminaires and sources appear to be more tolerable for insects. On the other hand, LEDs, especially cool white LEDs, have a higher blue content. This in turn attracts more insects, given that they react receptively to it (Fig. 3.9).

Nevertheless, the overall balance speaks for the LED. In this context, reference is often made to the studies carried out by Huemer et al. (2010) or Eisenbeis and Eick (2011). In the experiments, different types of lamps were tested to rate their attractiveness for insects. The studies showed that LED lamps attract the least insects.

However, the experiments cannot be regarded as sound evidence. In the tests carried out by Huemer et al. it is mentioned that in terms of luminous flux, luminous intensity and radiation the tests aimed to apply the same photometric values, but proof of this is missing throughout the study. There is neither information on the light distribution curve of the individual light sources nor information on the luminous flux. The luminous flux varies only according to the wattage of the luminaires (all luminaires have approximately the same wattage), but this does not say anything about the actual luminous flux of the light sources. Light sources of the same type, which differ only in their luminous colour, can have different luminous flux values. The experimental setup for the LED system is especially critical. Blanket values provided by the manufacturer were used. As a result, 25 W LEDs were used to replace "conventional" 80–100 W lamps. What "conventional" lamps were applied is not further explained. In the experiment it was also decided to use two LED light sources in each system. This meant that the LED test set-ups each had two light sources, whereas the other set-ups only had one.

In the experiment performed by Huemer et al. the conditions under which the different sources were tested were not consistent. The results are therefore virtually of no use. The statement that LEDs are less attractive to insects cannot be confirmed by this study.

The basic prerequisites for the tests on different luminaires carried out by Eisenbeis/Eick were also not consistent. In contrast to Huemer et al., in Eisenbeis'/Eick's experiment the illuminances of the individual lamp types were provided, measured at a height of approx. 1.5 m. The first differences between the lamp types were immediately evident. For example, the luminaire equipped with the high-pressure sodium vapour lamp had an illuminance of 42–56 lux and the LED modules tested showed an illuminance of only 12–16 lux. However, brightness is key to the degree of attractiveness of a light source. The brighter the light source, the more insects it is likely to attract (see Cleve 1967). A comparison between the light sources when the difference in illuminance is so high therefore leads to no conclusive results. In order to be able to make a meaningful comparison of the different lamp technologies, the illuminances of the different light sources should be similar.

That said, it makes even more sense to compare lamp technologies based on light sources with the same luminous flux. Illuminance is a variable factor, especially outdoors. The value can be quickly affected by a number of factors (e.g. the different phases of the moon, or wet surfaces). The luminous flux of the lamps, on the other hand, is a constant value, making it more suitable for comparative studies.

There are also differences in the light distribution curves of the luminaires; these differences also influence the degree of attractiveness of the light sources. A study that can be described as being on the right track in this regard was carried out by SRM Straßenbeleuchtung Rhein-Main GmbH (cf. Erfert 2012). The test examined the four most common lamp types: high-pressure sodium vapour (HST), metal halide (CDO), cool white LED, and warm white LED.

The various lamp technologies were applied in identical luminaire housings. Thus each lamp had a similar light distribution curve. Care was also taken to ensure that the lumen packages of the lamps were similar (high-pressure sodium vapour lamp 6600 lm, metal halide lamp 6700 lm). The LEDs, however, have a smaller lumen package (cool white LED 2720 lm, warm white LED 2720 lm), given that more powerful LED modules were not available when the study was carried out. In addition, the luminaires were only equipped with specific types of glass diffusers for the study: polymethyl methacrylate (PMMA) and polycarbonate (PC) were used. This is important because the covers influence the spectral composition of the light. For example PMMA does not allow UV radiation to pass through (Fig. 3.8, UV radiation <380 nm).

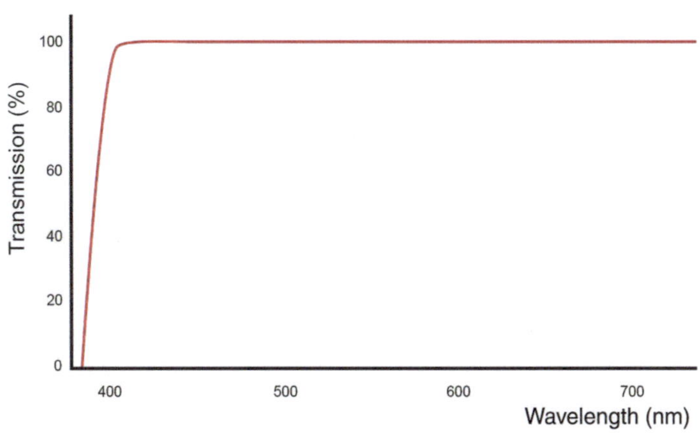

Fig. 3.8 Spectral PMMA transmission. (Source: Centre for Ion Beam Applications Singapore and Institute of Materials Research and Engineering Singapore, edited by the author)

The study has confirmed the findings from a number of other test studies:

- Insects are attracted most to metal halide lamps due to their UV and blue light components
- PMMA diffusers have a consistently positive effect on lamps with wavelength ranges below 400 nm, because they do not allow them to pass through
- Cool white LEDs are more attractive for insects than warm white LEDs. This can be explained by the higher blue component of the cool white LEDs (Fig. 3.9).

On the other hand, the study also shows that LEDs do not necessarily produce more insect-compatible light compared with other lamp technologies under practically the same conditions.

In Fig. 3.10 it can be seen that the difference between the approach rate of insects to high-pressure sodium vapour lamps and the cool white LEDs is only 15.6%. However, the luminous flux of the high-pressure sodium vapour lamp is 59% higher than that of the cool

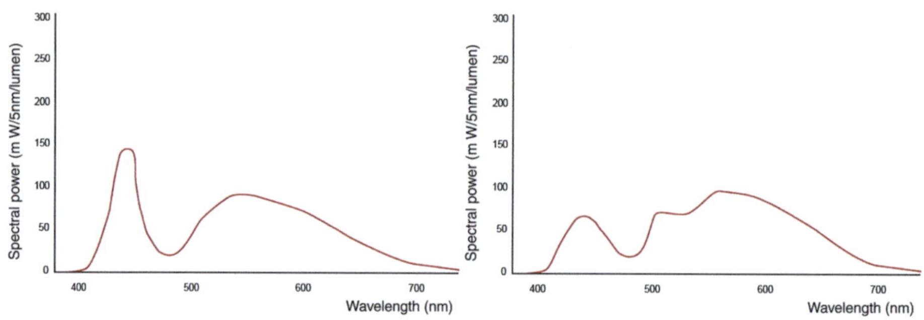

Fig. 3.9 Cool white LED (left) and warm white LED (right). (Source: LMK software)

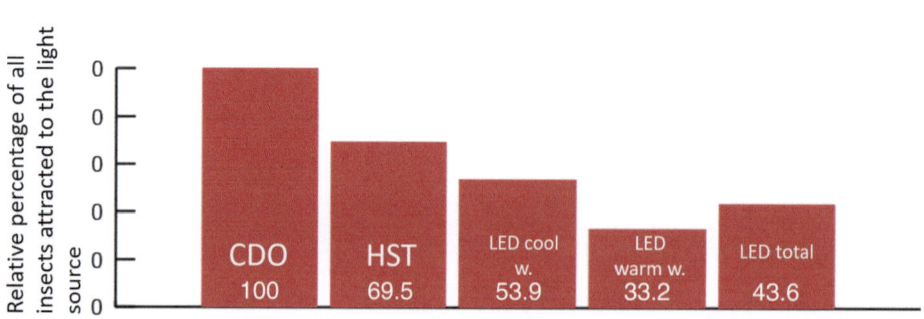

Fig. 3.10 Relative percentage of insects approaching light sources. (Source: Erfert 2012, edited by the author)

white LED. Given the same lumen package, it can therefore be assumed that the high-pressure sodium vapour lamp is even more compatible with insects than the LED.

In fact, there are tests that show that doubling the brightness increases the attraction potential by up to 70% (cf. Cleve 1967). The tests were made before the introduction of LEDs and therefore not with the LED spectrum. How high the attraction potential rises with the increase in the luminous flux of the LED cannot therefore be stated in concrete terms. Cool white LEDs would, however, be more "insect-incompatible" than high-pressure sodium vapour lamps even if the luminous flux were increased by 35%.

As a result of the study, a high-pressure sodium vapour lamp with a PMMA diffuser appears to be the most "insect-friendly" choice. Nevertheless, further research is required in the near future to classify LED technology correctly in terms of insect friendliness. According to current developments, almost all roads will be equipped with LED technology in the coming years. Reliable research findings relating to the degree of attraction LEDs have for insects would be desirable.

3.3.2 Effects of Sources of Attraction on Insect Behaviour

The attraction effect of light sources harbours several dangers for insects.

The heat generated by the luminaire head can cause burns if insects come into contact with it. Insects are killed directly or important body parts mutilated (cf. Kerstens 1961, pp. 119–138). Fast flying insects can also be killed or injured by flying into the luminaire housing (insects can reach speeds of 50–70 km/h (cf. Reinhardt and Harz 1996 and cf. Mazzucco 1952).

The design of the luminaire can also represent a potential hazard. If the luminaire head is not sealed properly, insects can find their way inside and not get out again. As a consequence, they will starve to death or burn (Fig. 3.11).

Non-flying insects which are attracted to light run the risk of being trodden on or run over on pavements or on the roadway. The generally high number of insects that accumulate around the luminaires also make for easy prey for their natural predators. This results in increased mortality (Malicky 1965, pp. 358–377).

Furthermore, during the period when they are attracted to the lights, insects cannot perform tasks that are important for their survival. This can lead to reproductive suppression or a reduction in the time spent on searching for food. The longer time the insects spend flying around the lights can also influence the activities mentioned, since they do not then have sufficient energy for other pursuits.

The light sources also attract insect populations from their original habitats to habitats that may not be suitable for survival.

In addition to the effects of the individual light sources, there are other negative effects caused by artificial lighting at night.

Fig. 3.11 Improperly sealed luminaire with dead insects. (Source: Johannes Käppler, unpublished)

3.3.3 Influence on the Circadian Rhythm

Nocturnal illumination also interferes with the circadian rhythm of insects. Certain activities pursued by insects are also controlled by the different levels of brightness throughout the day and night. For example, the luminous intensity must be below 1–10 lux in winter and 25 lux in summer for moths to awaken from their rest during the day (cf. Persson 1971, pp. 215–232). Conversely, diurnal insects require a similarly low luminous intensity to be able to rest and sleep at night. As can be seen from the studies of Cinzano (2002) and Cinzano et al. (2001) (parag. 3.1), this is not guaranteed in many urban areas. This shortens the activity time for nocturnal insects, whereas the activity time for day-active insects is longer. This again impacts reproduction and time spent searching for food.

Similar to birds, artificial lighting influences the time of hatching of insects. Thus the hatching time of butterfly pupae can be manipulated to a significant extent via artificial lighting. Beyond that, individual pupating creatures simply die off in the process. Creatures that hatch too early can have considerable problems finding partners and food. In addition, unacceptable weather conditions may prevail at the time of hatching (cf. Hayes et al. 1974, pp. 863–865).

It should not be forgotten that starry skies play an important role for the orientation of the insects. Studies undertaken show that dung beetles use the Milky way for orientation (cf. Dacke et al. 2013, pp. 298–300). Nocturnal illumination endangers this function.

The most important facts from this section are summarized in Table 3.4.

Table 3.4 Brief summary of light pollution effects on insects

Cause	Impact
Light source	• Attracted by light source, as a result: – Bulk of energy spent on flying – Reproductive suppression, less time for searching for food – Insects are attracted away from their natural environment – Killed or burnt when flying into the luminaire housing – Increased mortality rate when gathered around luminaire (easy prey)
LED spectrum (high blue component)	• Highly attractive to insects • Suspected attraction of new species of insects that are sensitive to shorter wavelength light
Artificial light at night (sky glow)	• Negative impact on the circadian rhythm • Interferes with hatching times • Orientation via the stars is undermined

Bibliography

Autrum, H./M. F. Bennett/B. Diehn/K. Hamdorf/M. Heisenberg/M. Järvilehto/P. Kunze/R. Menzel/ W. H. Miller/A. W. Snyder/D. G. Stavenga/M. Yoshida (Ed.), *Comparative Physiology and Evolution of Vision in Invertebrates* Berlin, Heidelberg 1979.

Barbur, J.L./A. J. Harlow/A. Sahraie, "Pupillary responses to stimulus structure, colour and movement", in: *Opthalmic and Physiological Optics* H. 12 (1992), P. 137–141.

Bartenbach, Christian/Walter Witting, *Handbuch für Lichtgestaltung. Lichttechnische und wahrnehmungspsychologische Grundlagen*, Vienna 2009.

Bartlett, Neil R./Clarence Henry Graham, *Vision and visual perception* New York et al. 1965.

Bass, Joseph/Fred W. Turek, "Sleepless in America: a pathway to obesity and the metabolic syndrome?", in: *Archives of internal medicine* Vol. 165, H. 1 (2005), pp. 15–16.

Bauer, R., *Untersuchung zur Anlockung von nachtaktiven Insekten durch Beleuchtungseinrichtungen.* Diploma thesis, Konstanz 1993.

Berthold, P., *Orientation in Birds* Basel, 1991.

Birkhead, T. R./Katrina van Grouw, *Die Sinne der Vögel oder Wie es ist, ein Vogel zu sein*, Berlin et al. 2015.

Böttcher, Marita (Ed.), *Auswirkungen von Fremdlicht auf die Fauna im Rahmen von Eingriffen in Natur und Landschaft. Analyse, Inhalte, Defizite und Lösungsmöglichkeiten; presentations and Results of the symposium of the same name on the island of Vilm from 06 to 09. December 1999*, Bonn, Bad Godesberg 2001.

Brugger, P./W. Marketl/M. Herold, "Impaired nocturnal secretion of melatonin in coronary heart disease", in: *The Lancet* Vol. 345, H. 8962 (1995), p. 1408.

Bullough, J. D./J. van Derlofske/C. R. Fay/P. A. Dee, "Discomfort glare from headlamps. Interactions among spectrum, control of gaze and background light level", in: *lighting technology* (2003), S. 21–25.

Bullough, John D./Mark S. Rea/Mariana G. Figueiro, "Of mice and women: light as a circadian stimulus in breast cancer research", in: *Cancer causes & control CCC* vol. 17, vol. 4 (2006), p. 375–383.

Bund Für Umwelt und Naturschutz Deutschland, Insektenfreundliche Leuchtmittel, 17.07.2015, (accessed: 10. 01.2017). http://www.bund-sh.de/themen_und_projekte/natur_und_artenschutz/naturschutz_in_der_gemeinde/beleuchtung/insektenfreundliche_leuchtmittel/

Cajochen, Christian/Mirjam Münch/Szymon Kobialka/Kurt Kräuchi/Roland Steiner/Peter Oelhafen/Selim Orgül/Anna Wirz-Justice, "High sensitivity of human melatonin, alertness, thermoregulation, and heart rate to short wavelength light", in: *The Journal of clinical endocrinology and metabolism* Vol. 90, H. 3 (2005), pp. 1311–1316.

Chen, Eric, "SeeingBlue", in: *nightscape* H. 80 (2010), P. 8–12.

Cinzano, P./F. Falchi/C. D. Elvidge, "The first World Atlas of the artificial night sky brightness", in: *Monthly Notices of the Royal Astronomical Society* Vol. 328, H. 3 (2001), pp. 689–707.

Cinzano, Pierantonio (Ed.), *Light pollution and the protection of the night environment. Proceedings of the conference Light pollution and the protection of the night environment Venice: let's save the night, held in Venice on May 3rd, 2002* Thiene 2002.

Cleve, K., "Münchner Entomologische Gesellschaft", in: *Nachrichtenblatt der Bayerischen Entomologen*, Vol. 16, H. 5/6 (1967), S. 33–53.

Creutz, Gerhard, *Geheimnisse des Vogelzuges* Wittenberg Lutherstadt 1989.

Da Silva, Arnaud/Mihai Valcu/Bart Kempenaers, "Light pollution alters the phenology of dawn and dusk singing in common European songbirds", in: *Philosophical transactions of the Royal Society of London. Series B, Biological sciences* Vol. 370, H. 1667 (2015), (accessed: 10. 01.2017). https://www.ncbi.nlm.nih.gov/pmc/articles/PMC4375366/

Dacke, Marie/Emily Baird/Marcus Byrne/Clarke H. Scholtz/Eric J. Warrant, "Dung beetles use the Milky Way for orientation", in: *Current biology* vol. 23, vol. 4 (2013), p. 298–300.

Dawson, A./V. M. King/G. E. Bentley/G. F. Ball, "Photoperiodic Control of Seasonality in Birds", in: *Journal of Biological Rhythms* Vol. 16, H. 4 (2001), p. 365–380.

Deutsches Institut für Normung, *Licht und Beleuchtung – Grundlegende Begriffe und Kriterien für die Festlegung von Anforderungen an die Beleuchtung*, Beuth Publishing House (DIN EN 12665:2011-09) 2011.

Dominoni, D.M./W. Goymann/B. Helm/J. Partecke, "Urban-like night illumination reduces melatonin release in European blackbirds (Turdus merula): implications of city life for biological timekeeping of songbirds", in: *Frontiers in Zoology* H. 10 (2013a), p. 1–10.

Dominoni, D.M./B. Helmet/M. Lehmann/H. B. Dowse/J. Partecke, "Clocks for the city: circadian differences between forest and city songbirds", in: *Proceedings. Biological Sciences/The Royal Society* vol. 280, vol. 1763 (2013b), (accessed: 10.01.2017). http://rspb.royalsocietypublishing.org/content/280/1763/20130593

Dominoni, Davide M./Jesko Partecke, "Does light pollution alter daylength? A test using light loggers on free-ranging European blackbirds (Turdus merula)", in: *Philosophical transactions of the Royal Society of London. Series B, Biological sciences* Vol. 370, H. 1667 (2015), (accessed: 10. 01.2017). http://rstb.royalsocietypublishing.org/content/370/1667/20140118

Dominoni, Davide M./Michael Quetting/Jesko Partecke, "Long-term effects of chronic light pollution on seasonal functions of European blackbirds (Turdus merula)", in: *PLOS ONE* vol. 8, vol. 12 (2013c), p. 1–9.

Eisenbeis, Gerhard, Klaus Fick, *Studie zur Anziehung nachtaktiver Insekten an die Straßenbeleuchtung unter Einbeziehung von LEDs*, in: Natur und Landschaft, Vol. 86, Vol. 7 (2011).

Erfert, Thomas, *Wirkung moderner Straßenbeleuchtung auf das Anflugverhalten von Insekten unter Berücksichtigung lichttechnischer Parameter*, 2012. (unpublished)

Esser, Karl, *Progress in botany* Berlin 2007.

Evans, William R./Yukio Akashi/Naomi S. Altman/Albert M. Manville, "Response of night-migrating songbirds in cloud to colored and flashing light", in: *North American Birds* Vol. 60, H. 4 (2007), pp. 476–488.

Fischman, A. J./A. J. Kastin/M. V. Graf/R. L. Moldov, "Constant light and dark affect the circadian rhythm of the hypothalamic-pituitary-adrenal axis", in: *neuroendocrinology* Vol. 47, H. 4 (1988), p. 309–316.

Fördergemeinschaft Gutes Licht, "Wirkung des Lichts auf den Menschen", in: *licht.wissen*, Vol. 19 (2014).

Frank, K., "Impact of outdoor lighting on moths: an assessment", in: *Journal of the Lepidopterists' Society* Vol. 42, H. 2 (1988), pp. 63–93.

Glickman, Gena/Robert Levin/George C. Brainard, "Ocular input for human melatonin regulation: relevance to breast cancer", in: *Neuro endocrinology letters* 23 Suppl 2 (2002), pp. 17–22.

Gwinner, Eberhard, Circannual clocks in avian reproduction and migration, in: *ibis* Vol. 138, H. 4 (1996), p. 47–63.

Hansen, Johnni, Increased Breast Cancer Risk among Women Who Work Predominantly at Night, in: *epidemiology* Vol. 12, H. 1 (2001), pp. 74–77.

Haupt, H., "Post-Tower" and Birds, March 2008, (accessed: 10. 01.2017). http://www.lichtverschmutzung.de/dokumente/PostTower_und_Vogelwelt.pdf

House, Erhard/Michael Smolensky, "Biological clocks and shift work: circadian dysregulation and potential long-term effects", in: *Cancer causes & control CCC* vol. 17, vol. 4 (2006), p. 489–500.

Hayes, D.K./B. Mr. Cawley/W. N. Sullivan/V. E. Adler/M. S. Schechter, "The Effect of Added Light Pulses on Overwintering and Diapause, Under Natural Light and Temperature Conditions, of Four Species of Lepidoptera", in: *Environmental Entomology* Vol. 3, H. 5 (1974), pp. 863–865.

Henke, H., *Empfehlungen für eine umweltverträgliche Außenbeleuchtung*, 1991.

Hentschel, Hans-Jürgen, *Licht und Beleuchtung. Theorie und Praxis der Lichttechnik*, 4th, reprocessed edition, Heidelberg 1994.

Hickman, Cleveland P./Wolf-Michael Weber, *Zoologie* 13th, updated edition, Munich, Boston et al. 2008.

Huemer, Peter, Hannes Kühtreiber, Gerhard Tarmann, Anlockwirkung moderner Leuchtmittel auf nachtaktive Insekten – Results of a field study in Tyrol, December 2010, (accessed: 10. 01.2017). http://www.hellenot.org/fileadmin/user_upload/PDF/WeiterInfos/10_AnlockwirkungInsektenFeldstudie_TLMFundLUA.pdf

Institut für Arbeitsschutz der Deutschen Unfallversicherung, Blendung - Theoretischer Hintergrund, May 2010, (accessed: 10. 01.2017). http://www.dguv.de/medien/ifa/de/fac/strahl/pdf/blendung_theorie.pdf

International Dark-Sky Association, Visibility, Environmental, and Astronomical Issues associated with Blue-Rich White Outdoor Lighting, May 2010, (accessed 10.01.2017). http://darksky.org/wp-content/uploads/bsk-pdf-manager/8_IDA-BLUE-RICH-LIGHT-WHITE-PAPER.PDF

Johnston, David W./T. P. Haines, "Analysis of Bird Mortality in October, 1954", in: *The Auk* Vol. 74, H. 4 (1957), p. 447–458.

Kempenaers, Bart/Pernilla Borgström/Peter Loës/Emmi Schlicht/Mihai Valcu, "Artificial night lighting affects dawn song, extra-pair siring success, and lay date in songbirds", in: *Current biology CB* Vol. 20, H. 19 (2010), pp. 1735–1739.

Kerstens, G., "Coleopterologisches vom Lichtfang," in: *Entomologische Blätter* Vol. 57 (1961), P. 119–138.

Kloog, Itai/Abraham Haim/Richard G. Stevens/Micha Barchana/Boris A. Portnov, "Light at night co-distributes with incident breast but not lung cancer in the female population of Israel", in: *Chronobiology international* Vol. 25, H. 1 (2008), pp. 65–81.

Kurtze, W., "Synökologische und experimentelle Untersuchungen zur Nachtaktivität von Insekten", in: *Zoologische Jahrbücher*, Vol. 101 (1974), pp. 297–344.

Lambrechts, M. M./J. Blonde/M. Maistre/P. Perret, "A single response mechanism is responsible for evolutionary adaptive variation in a bird's laying date", in: *Proceedings of the National Academy of Sciences of the United States of America* Vol. 94, H. 10 (1997), p. 5153–5155.

Leproult, R./E. F. Colecchia/M. L'Hermite-Balériaux/E. van Cauter, "Transition from dim to bright light in the morning induces an immediate elevation or cortisol levels", in: *The Journal of clinical endocrinology and metabolism* Vol. 86, H. 1 (2001), pp. 151–157.

Malicky, Hans, "Freilandversuche an Lepidopterenpopulationen mit Hilfe der JERMYschen Lichtfalle, mit Diskussion biozönologischer Gesichtspunkte", in: *Zeitschrift für Angewandte Entomologie*, Vol. 56, H. 1-4 (1965), pp. 358–377.

Mazzucco, K., *Rundschreiben der Österreichischen Forschungszentrale für Schmetterlingswanderungen No. 3*, Salzburg 1952.

Mustonen, A. M./P. Nobody/H. Hyvärinen, "Effects of continuous light and melatonin treatment on energy metabolism of the rat", in: *Journal of endocrinological investigation* H. 25 (2002), P. 716–723.

Narisada, Kohei/Duco Schreuder, *Light Pollution Handbook* Dordrecht 2004.

Naturwissenschaftlicher Verein Wuppertal, Mondorientiertes Flugverhalten - NVW, 16.07.2015, (accessed: 10. 01.2017). http://www.naturwissenschaftlicher-verein-wuppertal.de/sektionen/Entomologie/Zusatzinfos/mondorientiertes-flugverhalten

Noyes, Janet M., *Designing for humans* East Sussex, New York, 2001.

Partecke, Jesko/Thomas J. Van't Hof/Eberhard Gwinner, "Underlying physiological control of reproduction in urban and forest-dwelling European blackbirds Turdus merula", in: *Journal of Avian Biology* Vol. 36, H. 4 (2005), pp. 295–305.

Persson, Bert, "Influence of Light on Flight Activity of Noctuids (Lepidoptera) in South Sweden", in: *Insect Systematics & Evolution* Vol. 2, H. 3 (1971), pp. 215–232.

Philips Lighting Academy, Grundlagen von Licht und Beleuchtung 2008, (accessed 10.01.2017). http://images.philips.com/is/content/PhilipsConsumer/PDFDownloads/Germany/ODLI20150304_001-UPDde_DE-Grundlagen_von_Licht_und_Beleuchtung.pdf

Poot, H./J. E. Bruno/H. de Vries/M. Donners/M. R. Wernand/J. M. Marquenie, "Green Light for Nocturnally Migrating Birds", in: *Ecology and Society*, H. 2 (2008).

Reidenbach, Hans-Dieter/Klaus Dollinger/Günter Ott/Marco Janssen/Martin Brose, *Blendung durch optische Strahlungsquellen. Forschung Projekt F 2185*, Dortmund, Berlin, Dresden 2008.

Reinhardt, Rolf/Kurt Harz, *Wandernde Schwärmerarten (Totenkopf-, Winden-, Oleander- und Linienschwärmer)*, 2nd, unchanged. Ed. of the 1st ed. of 1989, Magdeburg, Heidelberg, Berlin, Oxford 1996.

Rich, Catherine/Travis Longcore, *Ecological consequences of artificial night lighting* Washington, DC 2006.

Robbins, K.R./A. A. Adekunmisi/H. V. Shirley, "The effect of light regime on growth and pattern of body fat accretion of broiler chickens", in: *growth* Vol. 48, H. 3 (1984), pp. 269–277.

Rodríguez, María I./Miguel Carretero/Germaine Escames/Luis C. López/María D. Maldonado/Dun-Xian Tan/Russel J. Reiter/Darío Acuña-Castroviejo, "Chronic melatonin treatment prevents age-dependent cardiac mitochondrial dysfunction in senescence-accelerated mice", in: *Free radical research*vol. 41, vol. 1 (2007), p. 15–24.

Rowan, W., "Relation of Light to Bird Migration and Developmental Changes", in: *nature* vol. 115 (1925), p. 494–495.

Salvendy, Gavriel, *Handbook of human factors and ergonomics* 4th ed., Hoboken 2012.

Schlag, Bernhard/Ina Petermann/Christoph Schulze/Gert Weller, *Mehr Licht – mehr Sicht – mehr Sicherheit? Zur Wirkung verbesserter Licht- und Sichtbedingungen auf das Fahrerverhalten*, Wiesbaden 2009.

Schmaljohann, Heiko/Felix Liechti/Bruno Bruderer, "Songbird migration across the Sahara: the non-stop hypothesis rejected!", in: *Proceedings. Biological Sciences/The Royal Society* vol. 274, vol. 1610 (2007), p. 735–739, (accessed: 10.01.2017). https://www.ncbi.nlm.nih.gov/pmc/articles/PMC2197203/

Schneider, T./H.-P. Thalau/P. Semm/W. Wiltschko, "Melatonin is crucial for the migratory orientation of pied flycatchers (Ficedula hypoleuca pallas)", in: *The Journal of Experimental Biology* (1994), S. 255–262.

Schoolmaster, K./M. Weber/W. Bogner/E. Schernhammer (Ed.), Application of melatonin action spectra on practical lighting issues, in: *Final report. The Fifth International LRO Lighting Research Symposium, Light and Human Health, November 3–5, 2002 Report No. 1009370* Palo Alto, CA 2004.

Steck, Bernhard, *For exposure of outdoor lighting systems to nocturnal insects. opinion* Berlin, 1997.

Stevens, Richard G./David E. Blask/George C. Brainard/Johnni Hansen/Steven W. Lockley/Ignacio Provencio/Mark S. Rea/Leslie Reinlib, "Meeting report: the role of environmental lighting and circadian disruption in cancer and other diseases", in: *Environmental health perspectives* vol. 115, vol. 9 (2007), p. 1357–1362.

Tamarkin, L./M. Cohen/D. Roselle/C. Reichert/M. Lippman/B. Chabner, "Melatonin Inhibition and Pinealectomy Enhancement of 7,12-Dimethylbenz(a)anthracene-induced Mammary Tumors in the Rat", in: *The Journal of Cancer Research* H. 41 (1981), P. 4432–4436.

Vaticon, M.D./C. Fernandez-Galaz/A. Esquifino/A. Tejero/E. Aguilar, "Effects of Constant Light on Prolactin Secretion in Adult Female Rats", in: *Hormones Research* Vol. 12, H. 5 (1980), pp. 277–288.

Wright, K. P./R. J. Hughes/R. E. Kronauer/D. J. Dijk/C. A. Czeisler, "Intrinsic near-24-h pacemaker period determines limits of circadian entrainment to a weak synchronizer in humans", in: *Proceedings of the National Academy of Sciences of the United States of America* Vol. 98, H. 24 (2001), p. 14027–14032.

Yoshizawa, Tōru, *Handbook of optical metrology. Principles and applications* Second edition, Boca Raton 2015.

International Comparative Analysis of Sources of Light Pollution in Urban Areas

4

4.1 Test Methodology and Implementation of the Comparative Analysis

In the context of the overview provided so far regarding the impact and effects of light pollution on various organisms, this chapter analyses the main sources and causes of light pollution in large cities. The results are findings that will contribute to solving the light pollution problem.

The cities selected for the comparative analysis were New York City, Boston, Toronto, Warsaw and Hanover. Given their different sizes and urban structures, they are representative of similar metropolises around the world.

North American metropolises with a distinct skyline	New York City, Toronto
Medium-sized North American towns/cities	Boston
European cities	Warsaw
Medium-sized European towns/cities	Hanover

International analysis also makes it possible to take into account the cultural differences that may influence lighting and light pollution.

Many previous analyses of light pollution use satellite imagery to view cities or areas. For example, Cinzano (2002) and Cinzano et al. (2001) (see Sect. 3.1), where satellites were equipped with special cameras so that nocturnal illumination could be better detected in certain regions. However, often only conventional satellite images of different areas are created and qualitatively compared with each other (Fig. 4.1).

These methods are relatively inaccurate, and the satellite perspective is not suitable for identifying light pollution sources in urban areas. The satellite perspective only enables the lighting of horizontal surfaces to be perceived. Vertical surfaces, such as facades, are not detected. This means that the photographs only reveal the road networks of the

E. E. Goronczy, *Light Pollution in Metropolises*,
https://doi.org/10.1007/978-3-658-29723-7_4

Fig. 4.1 Conventional satellite imagery, Chicago. (Source: NASA)

metropolises. What light (or source of light) the road surfaces reflect into the sky exactly, remains in the dark. With regard to assertions made about large regions (whole countries), and how much light is emitted into the atmosphere, however, these images are good and sufficient.

In order to be able to observe and compare the sources of light pollution in urban space, the study focuses on the heart of the urban space, where most of the action is. It is recorded from the human perspective, making it possible to detect and record light emissions in detail.

The main focus of the analysis and the comparison of the cities are the lighting hotspots in the respective metropolises. The different light emissions can be recorded very well at these locations. The lighting hotspots in the respective cities were determined beforehand on the basis of satellite images (Fig. 4.2). These areas form the regions in which photos were taken to determine luminance levels. Following further research using Google Streetview, at least ten main roads (main roads with a high volume of traffic, key transport routes, and roadways of economic importance) and ten side roads (roads with less traffic that are not absolutely necessary for direct access) were defined for the photo sessions. The photographs are labelled, quoting the number of the photograph, the abbreviation H for Hauptstraße (English: main road) or N for Nebenstraße (English: side road) and the initial letters of the city in which the photographs were taken (Fig. 4.3).

The selection consisted of more or less typical streetscapes with conventional luminaires in the respective areas. Subsequently, the urban space was divided into sections/categories.

Fig. 4.2 Lighting hotspot. (Source: NASA)

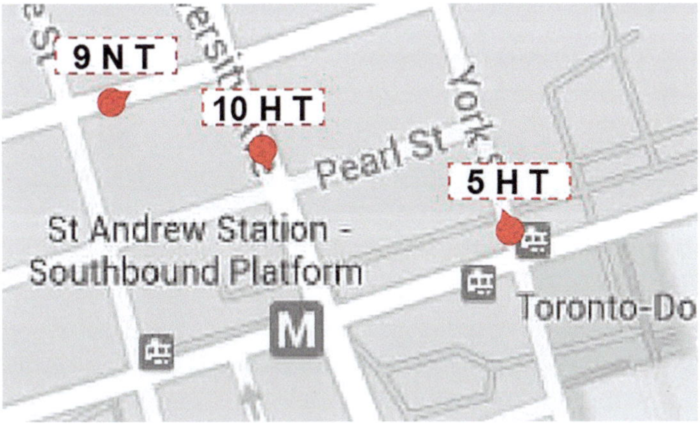

Fig. 4.3 Section of a map showing the photo session locations. (Source: Google Maps, edited by the author)

The main roads comprise the following: shop window/entrance, facade (including shop window/entrance), roadway and sidewalk/pavement. The side roads are divided into facade, roadway and sidewalk/pavement (Fig. 4.4).

Fig. 4.4 Image showing how the urban space is divided into the different sections

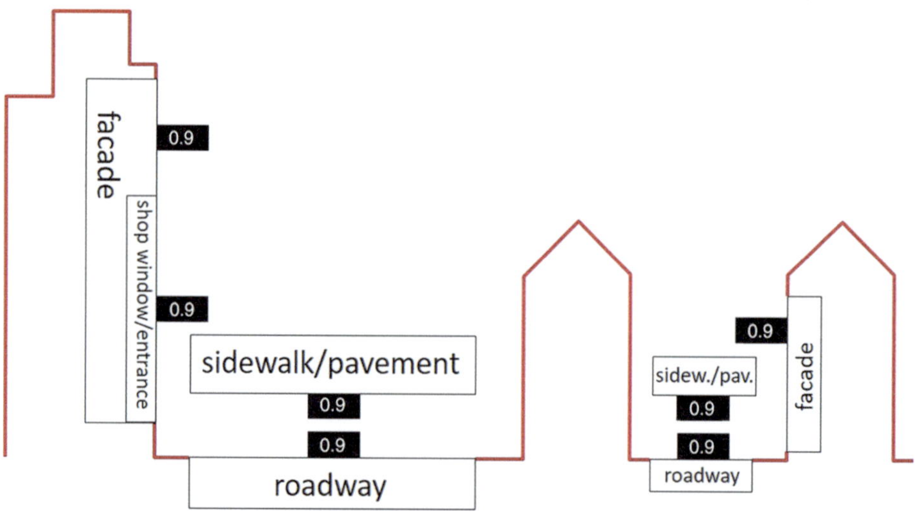

Fig. 4.5 Pictograph showing the different luminance levels in the urban space

The luminances were then recorded and the average for each section/category calculated and quoted in cd/m^2 (Fig. 4.5). The representative average is guaranteed by the number of photos.

The procedure makes it possible to compare the different cities in this regard.

The luminance of the streetlights was not measured in this analysis. Measuring the luminaire head is not useful in this case. Although luminaires generally have the highest luminance levels in urban areas, they do not emit light directly towards the sky/atmosphere (with the exception of non-cutoff luminaires). The light reaches the sky because it is reflected by the surfaces. It is the urban surfaces that the light from the street luminaires comes into contact with that need to be compared.

Given that not all sections/categories can be present in every photo, only the existing elements in any one photo were included in the calculation of averages.

Areas featuring extreme scenarios that deviate from the typical street scene, such as Central Park (dark compared to the urban environment) or the streets of Times Square (extremely bright compared to the urban environment) in New York City, were not included. These areas are an exception and would falsify the average value.

Further locations for photo sessions besides those previously selected could be defined during site inspection. All photographs of luminance levels were taken on site between 11.00 pm and 3.00 am. The photos were taken 1 h after sunset at the earliest to suppress the effect of the residual brightness of the setting sun. The photos of Warsaw were taken in April, of New York City, Boston and Toronto in May, and of Hanover in June 2015. All images were taken under dry weather conditions to avoid increased light reflections from wet surfaces. Photos were taken at a height of approximately 1.70–1.75 m, with the camera positioned between the lights. This prevented the light emitted by a luminaire positioned directly above the camera from distorting the results.

The camera used was a Canon EOS 550 D SLR with a Sigma 17–50 mm F2.8 EX DC OS HSM lens. This was used to shoot HDR images, which were converted into luminance photos using the "LMK Laboratory Software" program from TechnoTeam Bildverarbeitung GmbH, program version: 11.5.2. The program makes it possible to create a pixel-by-pixel (high-resolution) representation of the luminance based on the photos.

The sections on the metropolises comprise three subcategories with the following content:

- Preliminary analysis and expectations:
  Brief analysis of the urban area, special features, scenario to be expected on the basis of previous research, selection of locations for photo sessions (measuring points)
- Impressions during the walk through the town/city:
  Brief explanation of the activities pursued there, description of any special features, subjective impression of the urban lighting
- Evaluation of the measurement data:
  Objective evaluation of luminance photos, comparison (subjective perception), identification and classification of prime sources of light pollution.

Only relevant photos and data are shown in this book. The complete data sets and luminance images can be found in the online appendix. All photos have the same colour scaling (Fig. 4.6). The photos that served as a basis for the luminance images are available from the Institute for Building Services and Energie Design at TU Braunschweig (Institut für Gebäude- und Solartechnik).

A concluding summary and comparison of the metropolises, conclusions and findings from the data sets can be found in Sect. 4.7.

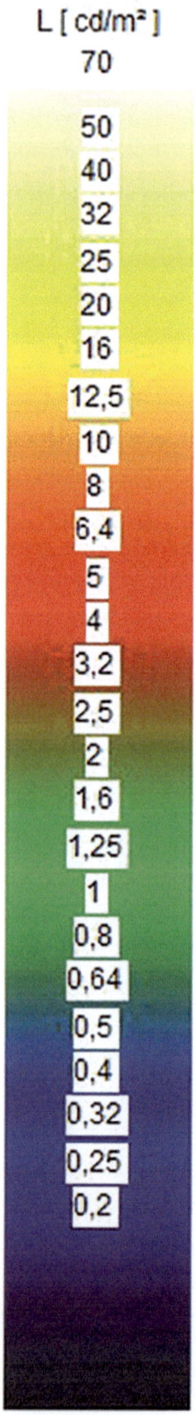

Fig. 4.6 Colour scaling for the luminance photos. (Source: LMK Laboratory Software)

4.2 New York City

4.2.1 Preliminary Analysis, Expectations

With over eight million inhabitants, New York City is by far the largest city in the analysis (see United States Census Bureau 2015). Like many major American cities, New York City has been designed in a regular grid structure.

The satellite image in Fig. 4.7 shows the lighting hotspot in close proximity to Central Park in Manhattan. The district stands out because of its high density of skyscrapers. The roads in this district usually consist of six lanes along the north-south axis, with the two outer lanes often used for parking. On the east-west axis, the roads generally consist of three lanes, mostly without parking. The individual carriageway width is approximately 3 m, resulting in a total road width of 9–18 m. On each side of the road there are sidewalks, which are around 3 m wide.

It can be assumed that, due to the higher number of lanes, the roads on the north-south axis are far busier traffic wise than those on the east-west axis. Accordingly, for the analysis the roads on the north-south axis are initially interpreted as main roads, and the roads on the east-west axis as side roads.

The study showed that the generously proportioned roads are illuminated by semi-cutoff luminaires (Fig. 4.8).

Fig. 4.7 New York City lighting hotspots. (Source: NASA, edited by the author)

Fig. 4.8 Typical street luminaire in NYC on site, luminaire type NYC A

In addition, the form of the luminaire suggests that the luminaires are equipped with discharge lamps. Furthermore, from the photo it can be speculated that the luminaires can be controlled individually. The module "perched" on top of the luminaire housing could well be a wireless control unit to enable this.

The luminaires are positioned along both sides of the road, although the height of the luminaires may vary in sections. No other luminaire types were discovered in the area under study.

The facades in the area consist largely of highly reflective materials such as glass, polished tiles or stone slabs (Fig. 4.9). The average building height is approximately 198 m (see Table A.1 in the appendix; explanation as to the calculations can be found in Chap. 5).

The area consists mainly of buildings with offices, hotels and shops on the lower floors.

It is expected that New York City has a severe light pollution problem. In a city that markets itself with the slogan "The City that never sleeps", it can be assumed that there is excessive use of light at night. Organisations such as Selene NY, which regularly submit draft laws to prevent light pollution, support this thesis (cf. Selene NY 2015).

The vertical structure of the city is interesting in terms of light emission technology. Due to the large number of tall buildings, New York features an unusually high surface area of building facades. It is likely that much of the light pollution stems from facades. Moreover, it can be assumed that any light entering the urban space is strongly reflected by the facade surfaces, resulting in a high overall level of brightness.

A new law dealing with the lighting of state-owned buildings, which was passed by Senator Carl L. Marcellino and signed by the Governor of New York, Andrew Cuomo, confirms this assumption (cf. Marcellino 2014).

The locations in the area where photos were taken of the representative streets are shown in Fig. 4.10.

Fig. 4.9 Photo showing typical facade surfaces in NYC

Fig. 4.10 Locations where photos were taken in New York City. (Source: Google Maps, edited by the author)

4.2.2 Impressions During the Site Inspection

During the site inspection, it was found that all streets in the area have a similarly high frequency of traffic. Thus the previously stipulated division into main and side roads in this area became obsolete. In order to be able to collect data relating to less busy roads, an area at the northern end of Central Park was included. In this case, the district was to represent the side roads. This is an area which consists mainly of living space and is therefore less frequented (Fig. 4.11). The satellite image (Fig. 4.7) shows that the illumination in this area features what was calculated as the average brightness found in the streets of New York.

Subjectively speaking, the main streets in New York have a relatively high level of brightness, which was actually assumed before any testing took place.

On site, however, street luminaires are not perceived as the main cause of the high brightness levels. The street luminaires are installed at regular intervals and deliver a uniform spread of light throughout the urban spaces without causing undue glare or being technologically overloaded. The luminaires are equipped with sodium vapour lamps and emit typically yellowish light.

The extreme brightness in this area seems to come mainly from the lower floors with their inherent commercial elements (representative entrances or shop windows), which are excessively lit and remain switched on even after midnight (Figs. 4.12 and 4.13).

Road traffic also contributes in a major way to high levels of brightness in urban areas. The main roads are heavily frequented by vehicles even late into the night (based on observations made from around 10.00 pm to 1.30 am). Especially at red traffic lights, when the vehicles come to a standstill, one can consequently perceive an increased brightness in the space.

These factors seem to be further enhanced by the facade surfaces with their highly reflective properties (Fig. 4.13). Interior lighting visible through windows in the facades, on the other hand, do not have a great influence on the overall brightness of the urban space.

It is also worth mentioning the illuminated construction sites in the area. They were perceived as the brightest source of light in the urban space (Fig. 4.14). These can be neglected to a large extent, since they are temporary installations.

Also worth mentioning is the visibility of Times Square within the urban realm. This is already perceivable from smaller streets several kilometres away thanks to the extreme brightness it generates (Fig. 4.15). Although this area appears to be important due to the high light emissions, Times Square was not part of this study for the reasons given in Sect. 4.1.

The "New York Street Lamp", information about which was researched in advance, is used for the entire area with one exception. The other luminaire type (luminaire type NYC B) is located in the W 57th St. (Fig. 4.16, location for photo session 2 H NY). It is also a semi-cutoff luminaire equipped with a sodium vapour lamp. The technical parameters appear to be identical to those of the other luminaires. There is no optical difference in the light distribution curve nor in the brightness gradient when viewed on site. The NYC B luminaire is also not perceived as causing glare.

Fig. 4.11 Additional locations where photos were taken in New York City. (Source: Google Maps, edited by the author)

Fig. 4.12 Location for photo session 3 H NY

The side streets in New York City differ structurally from the main streets mainly with regard to the height of the houses. Overall, the road space appears considerably darker compared to the main roads. This is probably due to the fact that the lower floors in this area are not used, and thus illuminated, for commercial or representative purposes. The other facade materials that appear here also play a role in terms of brightness in the urban space. The houses predominantly feature a natural or brick facade (Fig. 4.17), which reflects light to a lesser extent than, for example, glass facades.

In addition, the density of buildings in the area is lower. As a result, there are fewer surfaces that reflect light and therefore contribute to brightness levels.

The street widths of the main and side roads are identical. However, in the side streets the outer lanes are used for parking. The roadway itself is therefore smaller compared to the main roads.

Moreover, there are often rows of trees on the sidewalks in the side streets. The mounting height of the luminaire is above the treetops, so the areas under the trees are often unfavourably shaded. These areas are therefore less bright (Fig. 4.18).

Fig. 4.13 Location for photo session 21 H NY

Fig. 4.14 Construction site at the W 57th St

Fig. 4.15 Location for photo session 18 H NY

Fig. 4.16 Luminaire type NYC B

Fig. 4.17 Location for photo session 4 N NY

Fig. 4.18 Location for photo session 7 N NY

4.2.3 Evaluation of the Measurement Data

The measured data confirm the initial impression. Compared to the side roads, the main roads have consistently higher average values in the individual categories. This also makes them objectively brighter (Fig. 4.19).

As previously assumed, by far the most light is emitted in the lower floors on the main roads. At 5.3 cd/m^2, this value is almost eight times higher than the lowest average value. This value also stands out when considering the entire facade. Viewed as a whole, the average facade has a value of 1.9 cd/m^2, only the lower section of the facade reaching a value of 5.3 cd/m^2. This pronounced difference indicates that high light contrasts are very often at play (Fig. 4.20).

It is also interesting, however, that there is barely any difference between the measured data retrieved from the horizontal surfaces in the main and side streets (sidewalks, roadways). The horizontal surfaces are less present in the field of vision, which results from the nature of the human perspective. The fact that roadways and sidewalks in the main and side streets are practically equally bright was consequently not noticed when walking through the town. Human perception tends to focus on vertical surfaces. The facades, shop windows and entrances along the main roads feature extremely high values compared to the side streets. As a result, the urban spaces that make up the main roads appear to be many times brighter.

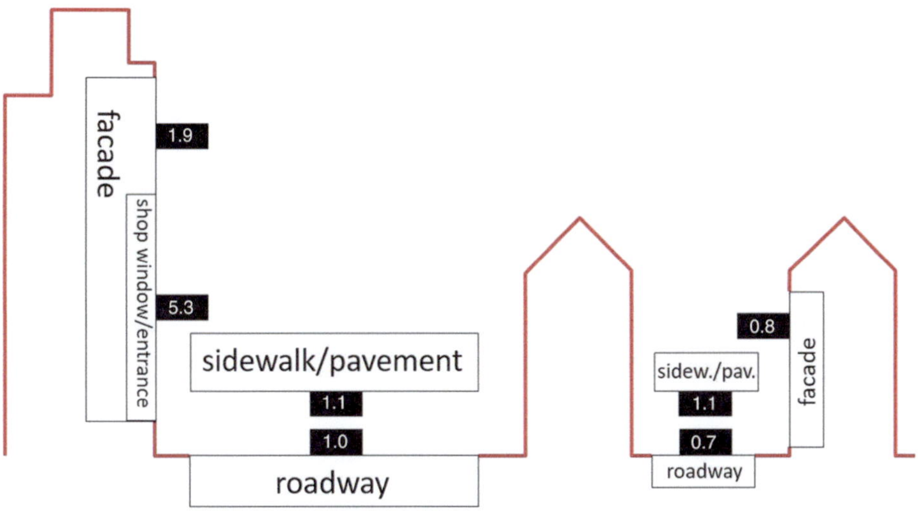

Fig. 4.19 Average luminance values (cd/m²) in the urban realm, New York City

Fig. 4.20 Location for photo session 3 H NY

Based on this conclusion, two facts can be assumed. On the one hand, it is plausible that most of the light emitted by the facades along the main roads is released into the atmosphere directly, and on the other hand, it is possible that a considerable amount of light is absorbed by the surfaces in the urban space. In both cases, this explains the marginal differences in the values.

It should be emphasised that values for horizontal surfaces at 0.7–1.1 cd/m^2 are not unusually high. In Germany, DIN 13201 stipulates a value of between 0.3 and 2.0 cd/m^2 for street lighting. Thus, even under strict German regulations, the roads would comply with the standards.

The luminance images also confirm that the reflections and backlit windows in the facades contribute to increasing the level of brightness in the urban space (Fig. 4.21).

Fig. 4.21 Location for photo session 24 H NY

Fig. 4.22 Location for photo session 9 N NY

At 5 cd/m^2, the windows located in the reddish area are close to the higher average value of the shop windows and entrances. Thus, contrary to speculations on site, they can indeed make a significant contribution to the overall level of brightness in the space.

When it comes to light reflected off facades, colour is a decisive factor. The brighter the colour, the more reflection takes place (Fig. 4.20, bright facade colour on the base of the building on the right). In the side streets, the houses usually have dark facades, so there is less reflection overall and the space is correspondingly darker compared to the main roads (cf. Figs. 4.22 and 4.20).

In the case of reflecting glass surfaces, bright objects reflected in the facade play a key role in the luminance of the facade. A reflecting facade can have a great influence, or very little influence, on the brightness of a space (Fig. 4.21).

It is also easy to see that the high density of buildings along the main streets alone means that there is more surface available to reflect the light, thus adding to the brightness in the space (cf. Figs. 4.22, 4.20, and 4.21).

The photograph taken at location 10 H NY is especially interesting (Fig. 4.23). In this case, the facade is fully illuminated using exterior projectors. In the New York City analysis, this was the building with the highest luminance value.

This shows that exterior facade lighting produces far more light in the urban space than illumination of the facade from the inside (cf. Figs. 4.23 and 4.21). The complete facade of

Fig. 4.23 Location for photo session 10 H NY

the building has an illuminance of approximately 8 cd/m^2 (Fig. 4.23), which exceeds the high average value of shop windows and entrances (Fig. 4.21).

The illumination of the urban space by road traffic cannot be analysed by means of the luminance images. The luminance images are superimposed HDR images, with a shutter speed of up to 20 s. This means that fast-moving luminous objects cannot be perceived in the space.

Following the site inspection and evaluation, one can say that in New York City the greatest potential for creating light pollution lies in the facade sections featuring shop windows and entrances that form part of the commercial areas.

4.3 Boston

4.3.1 Preliminary Analysis, Expectations

In the comparative analysis, Boston represents a medium-sized American city with around 600,000 inhabitants (cf. Emporis 2015a). The skyline is less pronounced compared to cities like New York, Los Angeles or Chicago. Due to the large number of highrisers, especially

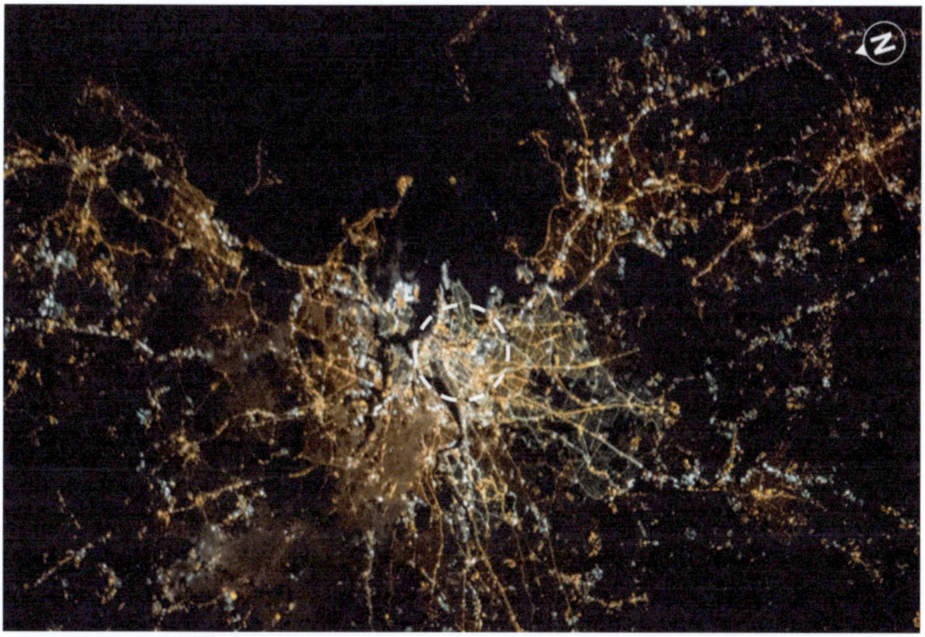

Fig. 4.24 Boston lighting hotspots. (Source: NASA, edited by the author)

in the centre, it still has on average higher buildings than most European cities (ibid.). Unlike many American cities, Boston has no grid structure. The streets and building areas have gradually developed naturally in the course of growth, without being subject to any principle of order. Thus the city has a structure that can be found in many European cities.

The satellite image shows that the metropolis has two lighting hotspots (Fig. 4.24). The area marked by the dotted circle is the first place with a high accumulation of light sources. The second lighting hotspot is located northeast of the marked circle. It marks the site of an airport, so it is not an urban area. Thus only the first lighting hotspot was analysed as part of the study.

The area where most of the brightness is produced is in the centre of the city. The highest buildings in Boston are also located in this area. On the basis of aerial photographs (Google Earth), it is possible to distinguish, without any doubt, the heavily frequented roads from the less frequented roads by the number of lanes. As a result, the main roads have a total of three or four lanes, with one lane mostly used as a parking lane. The side roads have two lanes. The carriageways are approximately 3 m wide and the sidewalks vary from between 4 and 5 m.

Fig. 4.25 Luminaire type B A, on site

Fig. 4.26 Luminaire type B B, on site

Fig. 4.27 Luminaire type B C, on site

The preliminary research showed that Boston has five different types of luminaire in the area under observation: luminaire type B A (Fig. 4.25), luminaire type B B (Fig. 4.26), luminaire type B C (Fig. 4.27), luminaire type B D (Fig. 4.28), luminaire type B E (Fig. 4.29).

Luminaire type B A is the most frequently represented. This luminaire type can be found in main streets as well as in side streets. This is a non-cut-off luminaire. The textured glass diffusers do not allow any judgement to be made about the lighting technology.

Fig. 4.28 Luminaire type B D, on site

Fig. 4.29 Lamp type B E. (Source: Google Earth)

Luminaires B B and B C are only to be found in main streets and are the second most frequently represented. Luminaire type B B is an LED luminaire. It can be assumed that these are full cutoff luminaires. A small module can be seen on the luminaire housing.

As observed in New York City, this could well be a wireless control unit. This suggests that the luminaires are controllable. Luminaire type B C is a semi-cutoff luminaire. It can be assumed that these luminaires are equipped with conventional discharge lamps.

Luminaire types B D and B E are only installed in side streets. Luminaire type B D is a luminaire with a "historical" design housing. The sources may well be LED substitutes. The upper and lower sections of the housing are made of glass. This suggests that it is a non-cutoff luminaire. The B E luminaire is a semi-cutoff luminaire. Given its shallow housing, it can be assumed that it is equipped with conventional discharge lamps.

Fig. 4.30 Facade surfaces Boston, on site

The facades in the area consist mainly of natural stone and brick. Some of the lower sections of the facades are darker and mostly made of wood (Fig. 4.30).

The area under analysis features a wide range of uses: bars, restaurants, offices, retail and residential areas. One area west of the centre is worth mentioning, since it comprises residential streets only without any retail spaces.

Given the large number of non-cutoff luminaires installed in Boston's urban areas, it can be expected that the city will have relatively high light emissions. The government's website shows that the metropolis is actively combating light pollution. Actions such as 'Lights out Boston', where in times of bird migration many lights are switched off (to facilitate the orientation of birds), show this (see City of Boston 2015a). The plan to successively replace all street lighting with LEDs also demonstrates the city's wish to combat light pollution (cf. City of Boston 2015b). It is particularly interesting to compare the already implemented LED street lighting with the conventional street lighting. It can be assumed that LED lighting generates less light in urban areas.

The locations for the photo sessions in roads in the area are shown in Fig. 4.31.

Fig. 4.31 Locations for photo sessions in Boston. (Source: Google Maps, edited by the author)

4.3.2 Impressions During the Site Inspection

A low overall brightness level can be observed on site. However, the brightness is sufficient to be able to perceive the surroundings to the full extent. The assumption made during the preliminary analysis that there would likely be significant sky glow in the city was not confirmed on site (Fig. 4.32).

Fig. 4.32 Location for photo session 7 H B

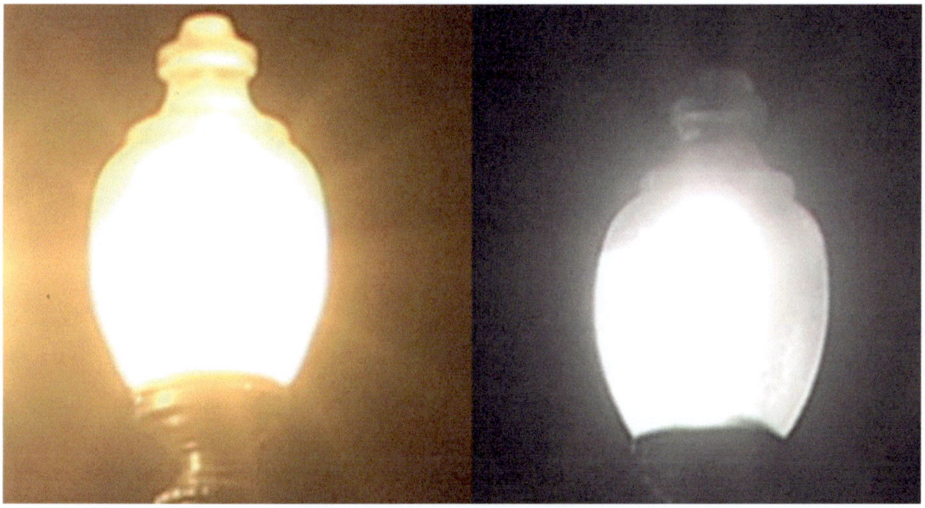

Fig. 4.33 Luminaire type B A, conventional (left) and LED (right)

Fig. 4.34 Close-up of the B A luminaire type

Street lighting is a special feature of Boston. As already explained in the preliminary analysis, the metropolis features a large variety of luminaire types. The conversion to LED technology expands the "product range". LED substitutes are gradually being installed in conventional luminaires. For example, luminaire type B A exists both with conventional sodium vapour lamps and with LEDs (Fig. 4.33).

With this type of luminaire, it should be emphasised that the LED version has a different light distribution curve than the conventional version. In contrast to the conventional luminaire, it directs light towards the street space (Fig. 4.34). The LED version is therefore not a non-cut-off luminaire in the conventional sense and will therefore emit less light into the atmosphere than the conventional version. When evaluating the measurement data, it is to be expected that the LED version therefore has better values with regard to light pollution.

Fig. 4.35 Detail of a photo taken at a chosen location 6 N B

Luminaire type B E is also available in an LED and sodium vapour version (Fig. 4.35). In this case it can be assumed that the luminaires do not differ in terms of the light distribution curve. In contrast to model B A, the luminaire already has a reflector module. One can strongly assume that only the discharge lamps with similar radiation characteristics to LEDs have been replaced by LED sources. As a result, the two versions have the same light distribution curve.

The LED luminaires differ from conventional luminaires in terms of colour temperature. They are "cooler". This creates a visually uneven overall picture in the city. This is reinforced by the fact that the LED versions appear to have a higher luminous flux. This also allows differences in brightness to be perceived on site.

The natural stone and brick facades generate practically no light reflections. The light is reflected softly and diffusely back into the urban space. With the exception of luminaire B B, there is no glare from the luminaires. The B B luminaire is an LED luminaire with lens optics. The individual LED circuit boards are thus visible. The glare is caused by the extremely high luminous flux from the comparatively small circuit board surfaces.

It should be noted that the side roads are frequently optically brighter than some of the main roads. One reason for this could be the narrower cross-section of the side roads. The side streets usually have the same lights as the main roads. However, because the facades are closer to each other and the road width is smaller, the light is absorbed less by the dark road surface. More light reaches the lighter coloured vertical surfaces. This results in a higher degree of reflection. The space becomes brighter (Fig. 4.36).

Also worth mentioning are the residential streets in the area with luminaire type B D installed (Fig. 4.37). They have a particularly low brightness level. The strong shading by

Fig. 4.36 Location for photo session 12 N B

Fig. 4.37 Location for photo session 8 N B

the trees also contributes to the fact that the space appears to be even darker. The level of brightness is nevertheless felt to be sufficient. The site inspection revealed that luminaire type B D has been retrofitted with LEDs.

4.3.3 Evaluation of the Measurement Data

Contrary to first impressions, the measurement data show that the urban space is not exceptionally dark (Figs. 4.38 and 4.39). The values even exceed all New York averages, with the exception of the shop window/entrance category. Nevertheless, the urban space in Boston was felt to be several times darker.

The feeling that some side roads in Boston appear brighter than the main roads is confirmed by the images (see photo session locations 7 H B, 1 N B, 2 N B, 3 N B, 12 N B in the appendix). The 'brightest' side road 3 N B (Fig. 4.40), for example, has a luminance on the facade that is 13 times higher than that in the 'darkest' main road 5 H B (Fig. 4.41). As mentioned in Sect. 4.3.2, this is due to the higher reflectance factor of the facade due to the narrower cross-section and the surface materials. As a rule, the facade surfaces in the side streets are lighter in colour, which reflects more light back into the space.

In addition, the image 1 N B (Fig. 4.42) demonstrates a classic case of clutter. Luminaires that have been installed too closely together lead to the space being "overexposed".

One factor that generates the low level of brightness in main road 5 H B is the difference in the building development on both sides of the road. One side of the road features a row of

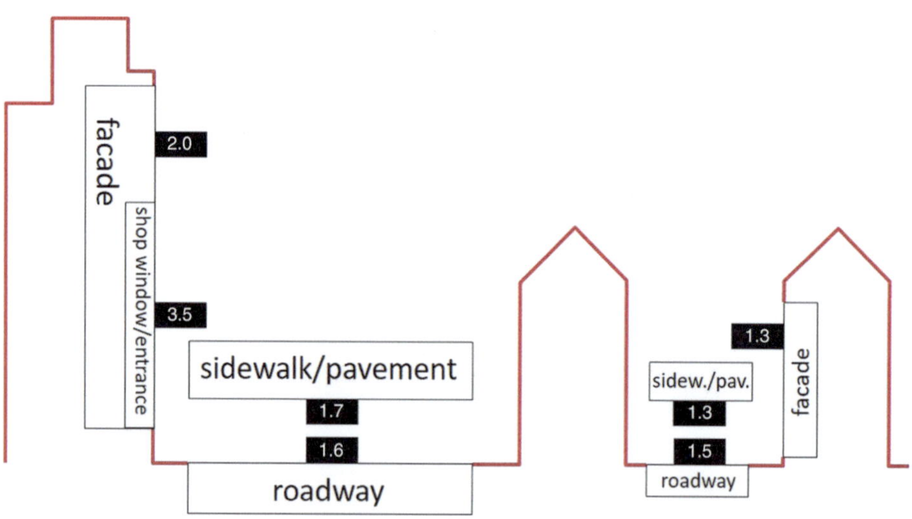

Fig. 4.38 Average luminance (cd/m^2) in the urban realm in Boston

Fig. 4.39 Location for photo session 7 H B

houses, and on the other side there is a park. Any light falling on the facades of the houses is reflected into the "emptiness" across the way (into the park) (Fig. 4.41).

The analysis of the measurement data collected for luminaire B A shows that there are no significant differences in brightness between the conventional luminaire equipped with a sodium vapour lamp and the LED version. In the photo taken at location 9 H B, both versions of luminaire B A can be seen in (Fig. 4.43). The lane directly next to the

Fig. 4.40 Location for photo session 3 N B

Fig. 4.41 Location for photo session 5 H B

Fig. 4.42 Location for photo session 1 N B

luminaires was measured, as were the luminaire heads from the photo session location. The luminance of the roadways was approximately 2.6 cd/m^2 for the sodium vapour lamps and approximately 3.0 cd/m^2 for the LED versions. The luminaire heads each had a value of approximately 45 cd/m^2. A similar findings were recorded for luminaire type B E.

In spite of the minimal deviations in the values, however, the urban space with the LED versions appeared brighter compared to the space with sodium vapour versions. This can be justified by the colour spectrum of the LED in conjunction with the human perception of brightness. Given their spectrum, LEDs tend to be perceived as brighter than sodium vapour lamps (see Sect. 3.1).

In image 3 H B it can be seen that the B B LED luminaire is a full cutoff lighting fixture. The lighting on the trees behind the luminaires shows no light emission above the luminaire housing (Fig. 4.44). Furthermore, the light is directed onto the road with very little spill light. This can be seen in the clearly defined surface on the roadway (red area). In this case, the space is almost exclusively lit via the reflections on the street. Without lower floors with brightly lit entrances to shops, the facades would have a luminance of only 0.4 cd/m^2. The reflections on the facades can therefore be classified as having a low light pollution potential.

The sections of roadway directly beneath the lights, on the other hand, has a luminance of approximately 4 cd/m^2. How much of this increased value is reflected into the atmosphere cannot be determined. Therefore, the overall balance of the road space cannot be classified.

Figure 4.45 shows that in comparison with the LED version of luminaire type B A, luminaire type B C emits more light onto the road and into the surroundings. On the one

Fig. 4.43 Location for photo session 9 H B

hand, this can result from the higher luminous flux of the luminaires and, on the other hand, from the different light distribution curves. A mixture of both factors is also plausible.

Luminaire type B D, which can be found in residential streets, has the lowest brightness level of all luminaire types with an average luminance of 8 cd/m^2. The measured data confirm the experienced impressions on site.

The light patterns on the wall in Fig. 4.46 show that this is a luminaire that emits part of the light upwards and most of the light downwards. Although light is also emitted upwards, it can be assumed that only a very small part reaches the atmosphere and thus little light pollution is produced. The upper section of the facades in Fig. 4.46 is only gently illuminated.

Fig. 4.44 Location for photo session 3 H B

The advantages of a combination of luminaires and trees can be seen in Fig. 4.37. The crown of the tree reflects and absorbs most of the light radiating upwards. The light is "trapped" between the street and the treetops. Thus, only a small portion of the light is emitted into the atmosphere, while at the same time the urban space between treetops and street is illuminated (Fig. 4.47).

Also worth mentioning is the photo taken at location 4 N B (Fig. 4.48), which shows a side road without street lights. Nevertheless, the road surface has a luminance of 1.12 cd/m^2. This value is only slightly lower than the average value in this category. The level of brightness in the street is generated through the brightness in the main streets and on the facades. The photograph raises the question of whether it is generally necessary to equip all streets in urban areas with lighting systems.

In conclusion, the highest luminance values were recorded in the shop windows/entrances category. However, it should not go unmentioned that there is also a great potential for light pollution from the side roads. Shop windows and entrances are only relevant in the city centre, where they can be found in abundance. The percentage of surface area of secondary roads is substantially higher, making them more relevant in terms of light pollution.

Fig. 4.45 Location for photo session 4 H B

Fig. 4.46 Location for photo session 9 N B

Fig. 4.47 Location for photo session 8 N B

Fig. 4.48 Location for photo session 4 N B

4.4 Toronto

4.4.1 Preliminary Analysis, Expectations

With approximately 5.5 million inhabitants, Toronto is the largest city in Canada. Similar to New York City, it has a distinct skyline with 419 official skyscrapers (see Emporis 2015c). Toronto's urban structure resembles that of New York City, where the city was also laid out in a grid. However, the Toronto grid is less strict, as is the case in New York. The dimensions of the sites often differ here. In addition, roads often exist that do not run axially to the grid, which disrupts the structure to a certain extent.

Toronto's largest lighting hotspot is in the centre of the skyline, in the Entertainment District (Fig. 4.49). The area is located on the southern shoreline of the city.

The roads defined in the analysis as main roads have between four to six lanes, each 3 m wide. In addition, the central zone of the main roads is also used for rail traffic in parts. This is approximately 9 m wide.

Unlike in New York City, the blocks of houses often have smaller streets that break through the block structure. These are defined as side roads for the analysis. The side roads have one or two lanes, each with a width of 3 m. The sidewalks along the main roads are

Fig. 4.49 Toronto's prime lighting hotspot (dotted area). (Source: NASA, edited by the author)

Fig. 4.50 Luminaire type T A, on site

generally approximately 3 m wide. No regular dimension can be detected for the side roads. Some sections have three-metre sidewalks, where other sections have no sidewalk at all.

The preliminary research showed that four different types of luminaires can be found in the area. Luminaire type T A, luminaire type T B, luminaire type T C, luminaire type T D (Figs. 4.50, 4.51, 4.52, and 4.53). Luminaire types T A and B are most common in this area.

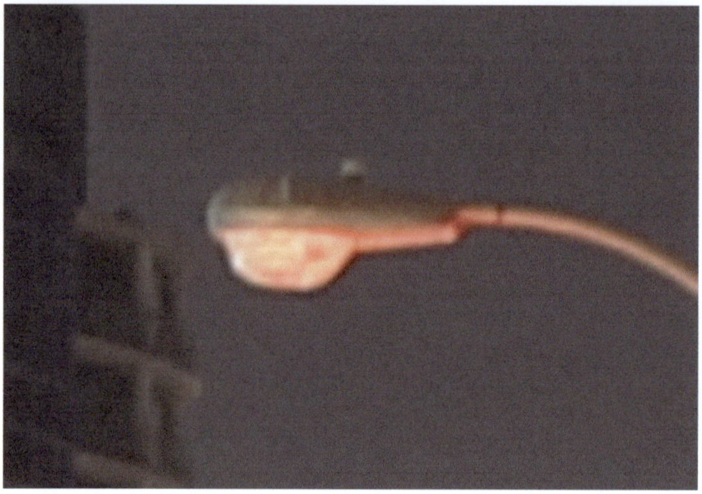

Fig. 4.51 Luminaire type T B, on site

Fig. 4.52 Luminaire type T C, on site

The mounting height of these two luminaires is twice as high (approximately 10 m) as that of the other luminaires. The luminaires feature semi-cut-off shielding. On the basis of the photographs, it can be assumed that these are equipped with discharge lamps. Similar to the luminaires in Boston and New York City, the module mounted on the luminaire housing of

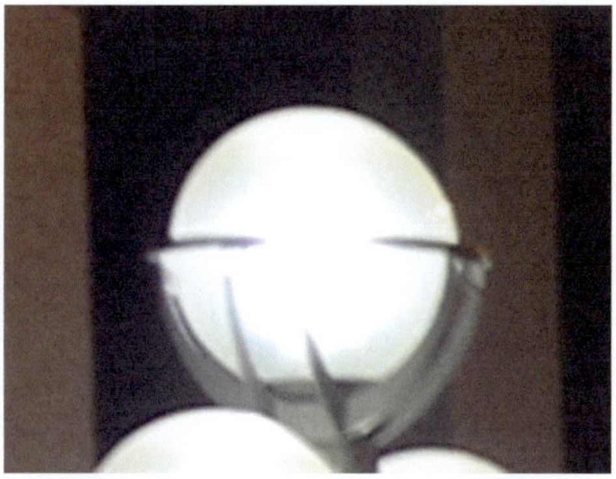

Fig. 4.53 Luminaire type T D, on site

luminaire type T B suggests that these luminaires can be controlled using wireless technology.

Luminaire types T C and T D are installed sporadically in the area; they have a lower mounting height (approximately 5 m) compared with the other luminaire types mentioned. Luminaire type T C is also a semi-cutoff luminaire, whereas luminaire T D is a non-cutoff luminaire and therefore emits its light in all directions. Here, too, it can be assumed that the luminaires are equipped with conventional discharge lamps.

There does not appear to be any clearly defined lighting layout in the area to be investigated. The luminaire types mentioned can be found in both main roads and side roads.

The urban realm in Toronto features a lot of reflective surfaces. The facades consist mainly of glass or highly polished facade panels (Fig. 4.54). Furthermore, the materials used for the facades are predominantly light in colour.

The area features office, catering, retail and residential space. There are also a number of nightclubs, bars and venues such as the Rogers Centre multi-purpose stadium and Ripley's Aquarium.

Especially in North America it is well known that Toronto is located unfavourably on a highly frequented migratory bird route. It is estimated that 1.5–2 billion migratory birds die annually in the United States and Canada as a result of human intervention in the ecosystem (FLAP CANADA 2015). A large proportion of the birds' deaths occur in Toronto, given its location. The FLAP CANADA organization, in particular, is committed to measures aimed at creating a more bird friendly environment in order to minimise the number of victims (ibid.).

Lighting recommendations for the City of Toronto have existed since 2011. On the one hand, they regulate illuminance levels in different zones, but on the other hand they also regulate outdoor lighting (cf. Hiscocks 2011). The extent to which these recommendations have been implemented did not become apparent in the preliminary research.

For the above reasons, it can be assumed that Toronto is very pro-active when it comes to combatting light pollution. Accordingly, low but sufficient luminances are expected. It

Fig. 4.54 Facade surfaces Toronto, on site

will be interesting to see how the metropolis intends to manage the situation with its highly reflective facades.

Figure 4.55 shows the locations pinpointed for the photo sessions in the representative streets in the area.

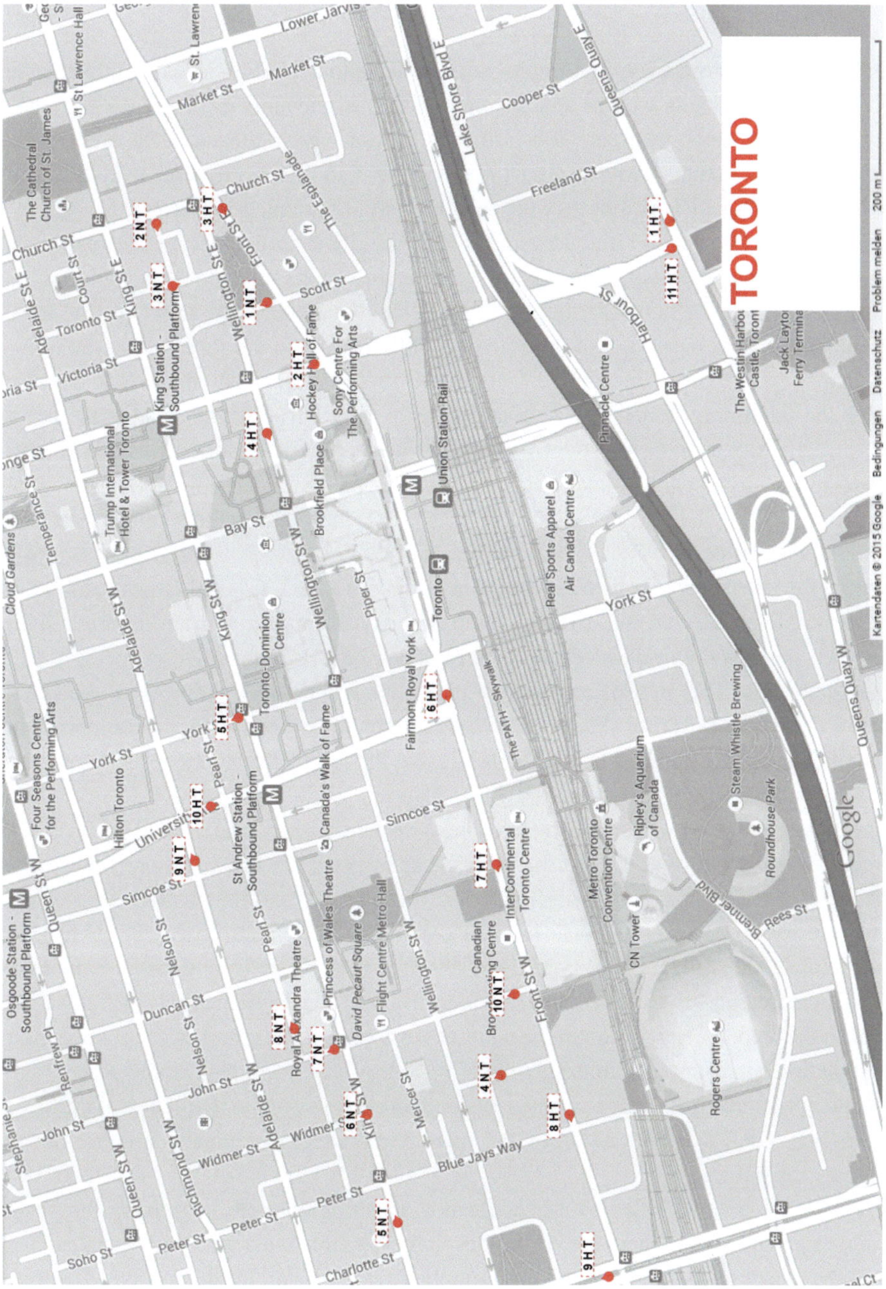

Fig. 4.55 Locations for photo sessions in Toronto. (Source: Google Maps, edited by the author)

4.4.2 Impressions During the Site Inspection

Compared to New York City, first impressions indicate that Toronto is less bright at street level. The urban space is well lit and also remarkably homogeneous. The facades have a low overall brightness, with the lower facade sections with the entrance areas being somewhat brighter (Fig. 4.56). Similar to New York City, there are representative commercial areas in this district, but they are less strongly illuminated in Toronto.

The urban space is mainly illuminated by the T A and T B luminaire types. As explained above, these have a high mounting height, which results in the uniform illumination of the urban space. On site the light distribution curve of the two models appeared to be the same.

In addition to the basic lighting provided by the two luminaire types, some buildings have their own lighting, which is designed more for decorative purposes. It is doubtful whether the lighting provided by these luminaires is necessary for the urban space per se (Fig. 4.57, the right side of the street without decorative luminaires is sufficiently bright). The two luminaire types T A and T B are each equipped with sodium vapour and metal halide lamps. The urban space appears brighter with the metal halide lamps than with sodium vapour lamps.

During the site inspection, it was noted that luminaire type T C is only installed in areas with low buildings. It is a sodium vapour lamp which delivers relatively strong bright light. The light is perceived as very direct or "sharply contoured".

Luminaire type T D is used in the area partly for functional lighting (Fig. 4.58) and partly for decorative purposes (Fig. 4.57). These luminaires were only found equipped with metal halide lamps. In contrast to the T C luminaire, the luminaire produces pleasantly soft light. This is the result of the structured glass diffuser.

One disadvantage, however, are the light spots on the horizontal surfaces. This gives rise to irregular patterns of light in the urban space, especially when the luminaires are spaced further apart.

The side roads in Toronto appear brighter compared to the main roads (Fig. 4.59). As already explained in the preliminary analysis, these feature the same luminaires as the main roads. The increased brightness is likely due to the fact that the narrower side streets are subject to the same design guidelines as the main streets. With a narrower cross-section, this means overexposure and consequently a higher level of brightness in the space.

Another point that has come to light is the large differences in the frequency of traffic on the main and side roads. Accordingly, the main roads were sporadically brighter due to increased road traffic.

The highly reflective facade surfaces do not play a major role in making the space brighter, as presupposed. There are no facades in the area that are actively illuminated. The light on the facades is basically spill light from the street lighting, traffic flow and windows lit from the inside (Fig. 4.60).

Consequently, there is little light in the space which can be reflected by the highly reflective surfaces.

Fig. 4.56 Location for photo session 11 H T

Fig. 4.57 Location for photo session 4 H T

Fig. 4.58 Location for photo session 10 N T

Fig. 4.59 Location for photo session 2 N T

Fig. 4.60 Location for photo session 5 H T

4.4.3 Evaluation of the Measurement Data

The measured data confirm the impression on site. Toronto's vertical surfaces have a lower brightness level than Boston or New York City. The urban space is not only subjectively perceived as being darker, but also objectively (metrologically proven) to be so, compared to the previous metropolises (Figs. 4.61 and 4.62).

The overall uniform impression gained of the urban space can also be confirmed by the measured data. There are only marginal differences between the values in the different categories. Only the shop window/entrance category stands out given its increased values (Fig. 4.61).

However, this was foreseeable and is in line with the experience gained on site. Shop windows and entrances have the function of attracting attention; in many places this is done by means of increased brightness so they stand out in their surroundings. The shop windows/entrances in Toronto also have the highest luminance levels.

With regard to luminaire types T A and T B, the luminance images cannot be used to determine whether the luminaire types feature the same or different light distribution curves.

However, it should be noted that there are differences between the luminaires equipped with sodium vapour sources and those equipped with metal halide lamps. The luminaires equipped with metal halide lamps deliver more uniform lighting and a lower brightness level (Fig. 4.63). However, the streets with metal halide lighting were perceived as being brighter during the site inspection. This can be explained by the light spectrum of metal halide lamps. Similar to LEDs, metal halide lamps have an increased blue-green

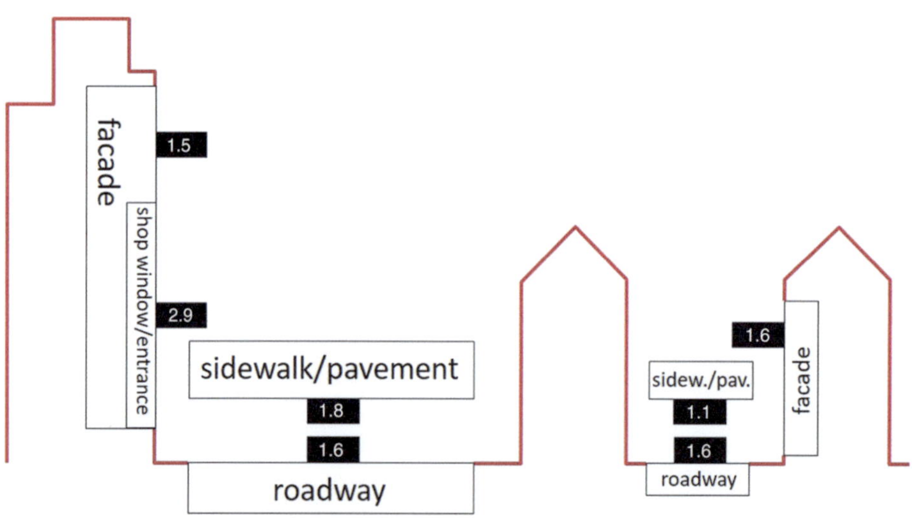

Fig. 4.61 Average luminance (cd/m^2) in the urban realm in Toronto

Fig. 4.62 Location for photo session 11 H T

component, to which the human eye is more sensitive. Urban spaces with a low measured value therefore may subjectively appear to be brighter (see Sect. 3.1).

The luminance images show light spots beneath the luminaires equipped with sodium vapour lamps. These light spots were not perceived during the site inspection (Fig. 4.64).

Fig. 4.63 Location for photo session 10 H T Metal steam equipment

Fig. 4.64 Location for photo session 8 H T. Sodium vapour lamps

Fig. 4.65 Location for photo session 4 H T

The difference in light distribution can be caused by the fact that the different light sources feature a different light distribution curve. Another reason may be the different reflectors in the luminaires, which also affect light distribution.

Figure 4.65 shows the T D luminaire installed directly in front of a facade. In this case, it serves as a decorative element. The luminance image shows that this results in clutter. The luminaires are installed in close proximity to one other, which results in disproportionately high overall lighting in this urban space.

The side streets were generally perceived to be brighter than the main roads during site inspection. This is confirmed by the luminance images. The side roads have an average value of 1.6 cd/m^2 and the main roads 1.5 cd/m^2. Although the value differs by only 0.1 cd/ m^2, it should be noted that the value calculated on the main street facades included the shop windows and entrances. Thus the actual values of the sections of facade above the first floors, where the shop windows and entrances are located, are substantially lower. Compared with the facade of the side streets, they are thus considerably darker (cf. Fig. 4.66 with Figs. 4.63 or 4.64).

One aspect revealed by the luminance photos is the strong asymmetrical light distribution of the T D luminaire. This was not consciously perceived during the site inspection. Figure 4.67 shows that the majority of the luminous flux is directed towards the sidewalk. The road surface receives very little light. This means that the luminaire does not feature typical non-cutoff light distribution, whereby the light is emitted into the space through

Fig. 4.66 Location for photo session 3 N T

Fig. 4.67 Location for photo session 10 N T

Fig. 4.68 Location for photo session 3 H T

360°. As a result, this type of luminaire will produce less light pollution than conventional non-cutoff luminaires.

The T C luminaire delivers relatively uniform lighting in the space, contrary to what one might feel (Fig. 4.68, right side of the street). The unpleasant lighting effects perceived cannot be evaluated using the luminance images. The images can only depict quantitative amounts of light, but not the quality or nature of the light generated.

Finally, it should be noted that very little of the brightness in the urban space results from the facades of the high-rise buildings (Fig. 4.62). Glare from the luminaires is also reduced to a minimum given that the light sources are positioned at a high level. The luminaires are well shielded from a distance. The closer one gets to the street luminaires, so they retreat from one's field of vision (given that they are mounted high above), which reduces the glare potential when viewing the overall street scenario. Given their low level of brightness, the highly reflective facade surfaces can be neglected. There is hardly any light to be reflected.

For its size, Toronto has low luminance levels. In the main, the light pollution problem comprises the side roads that are too bright plus the evidence of light clutter in isolated situations.

4.5 Warsaw

4.5.1 Preliminary Analysis, Expectations

Warsaw represents a major European city with 2.3 million inhabitants (cf. Emporis 2015d). Like many European cities, Warsaw has grown "naturally". The building plots and roads are not subject to a regulated structure, as is often the case in North America. The satellite photo (Fig. 4.69) shows the lighting hotspot in the centre of the city.

The area under investigation has a large number of different buildings. These include prefabricated buildings, old buildings and also several skyscrapers. The highest building in the area and the city is the Pałac Kultury i Nauki with a height of approximately 230 m (Fig. 4.70) (ibid.).

Preliminary research via Google Earth showed that Warsaw has a clearly structured, highly frequented road network. Smaller, less frequented streets branch off from the main roads. The main roads have two or three lanes in each direction, separated in the middle by a green strip, tram tracks, street lights and other urban elements (Fig. 4.71). The roadway is approximately 3 m wide. In addition to the road space, there are also some approximately four-metre wide sidewalks and three-metre deep parking bays on some stretches. These

Fig. 4.69 Warsaw lighting hotspots. (Source: NASA, edited by the author)

Fig. 4.70 Pałac Kultury i Nauki (left), on site

Fig. 4.71 Typical main road, on site

Fig. 4.72 Luminaire type W A, on site

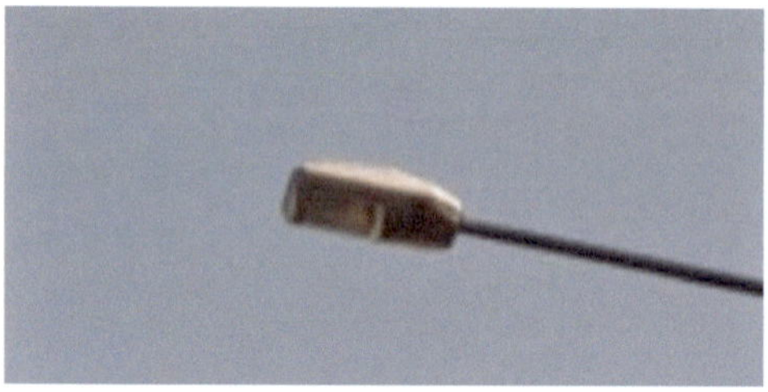

Fig. 4.73 Luminaire type W B, on site

streets are defined as main roads in the analysis. The smaller roads have between two to four lanes and have no central reservation. Here, too, the sidewalks are approximately 4 m wide. In addition, these roads feature three-metre wide parking bays. These roads are defined as side roads in the analysis.

The search carried out via the Internet revealed a total of four different types of luminaire. Luminaire type W A (Fig. 4.72), luminaire type W B (Fig. 4.73), luminaire type W C (Fig. 4.74) and luminaire type W D (Fig. 4.75).

Luminaire type W A is by far the most commonly applied in the area, followed by type W B. Luminaire type W B seems to be a further development of type W A. This luminaire can also be found in streets with W A type luminaires. It would appear that luminaire type W A is no longer being manufactured and that in the event of a defect it will be replaced by the W B luminaire. Very few W C luminaires have been installed in the area. The W D luminaires are installed at the location marked "Centrum", along the main road at Pałac Kultury i Nauki.

Luminaire type W D is actually a luminaire cluster, the overall luminaire comprising three to six individual luminaires (Fig. 4.75).

Fig. 4.74 Luminaire type W C, on site

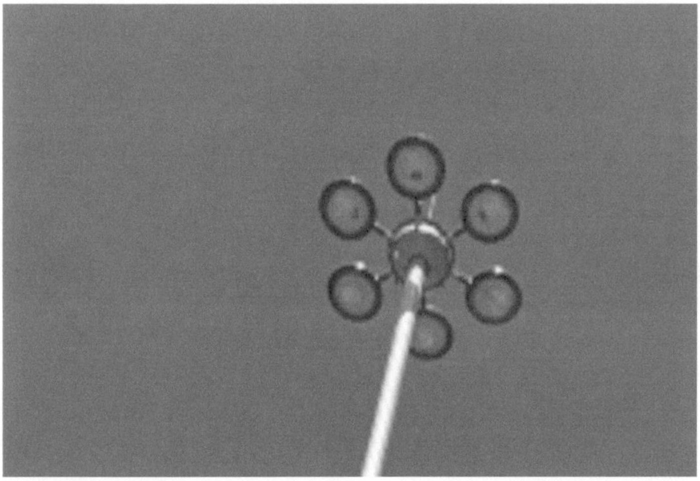

Fig. 4.75 Luminaire type W D, on site

In the area under study, there appears to be have been no clear system in place for the selection of luminaires.

The luminaires that are part of this study are semi-cutoff luminaires. Their form suggests that they were conventionally equipped with sodium vapour or metal halide lamps.

Fig. 4.76 Luminaire cluster made up of W A type luminaires, on site

The high number of luminaires in the urban realm is particularly striking, often with several luminaires mounted on one mast (independent of luminaire type) (Fig. 4.76).

Unless the luminaires have an exceptionally low luminous flux, extreme luminance levels can be expected from the clutter effect.

Other eye-catching features are the many billboards in the city on which luminaires are mounted. The advertising is illuminated at night, which means that a high level of light emission can also be expected here.

As in the cities described above, this area also has all usage profiles (restaurants, commercial buildings, residential housing).

The surfaces within the urban space feature predominantly dull-looking facade materials such as natural stone or plaster (Figs. 4.82 and 4.84).

In the preliminary research, no documents were found which indicate that lighting would be reduced to support "the prevention of light pollution". As mentioned above, clutter can give rise to very high levels of brightness.

Figures 4.77 and 4.78 show the locations selected for photo sessions in the representative streets in the area.

Fig. 4.77 Locations for photo sessions in Warsaw. (Source: Google Maps, edited by the author)

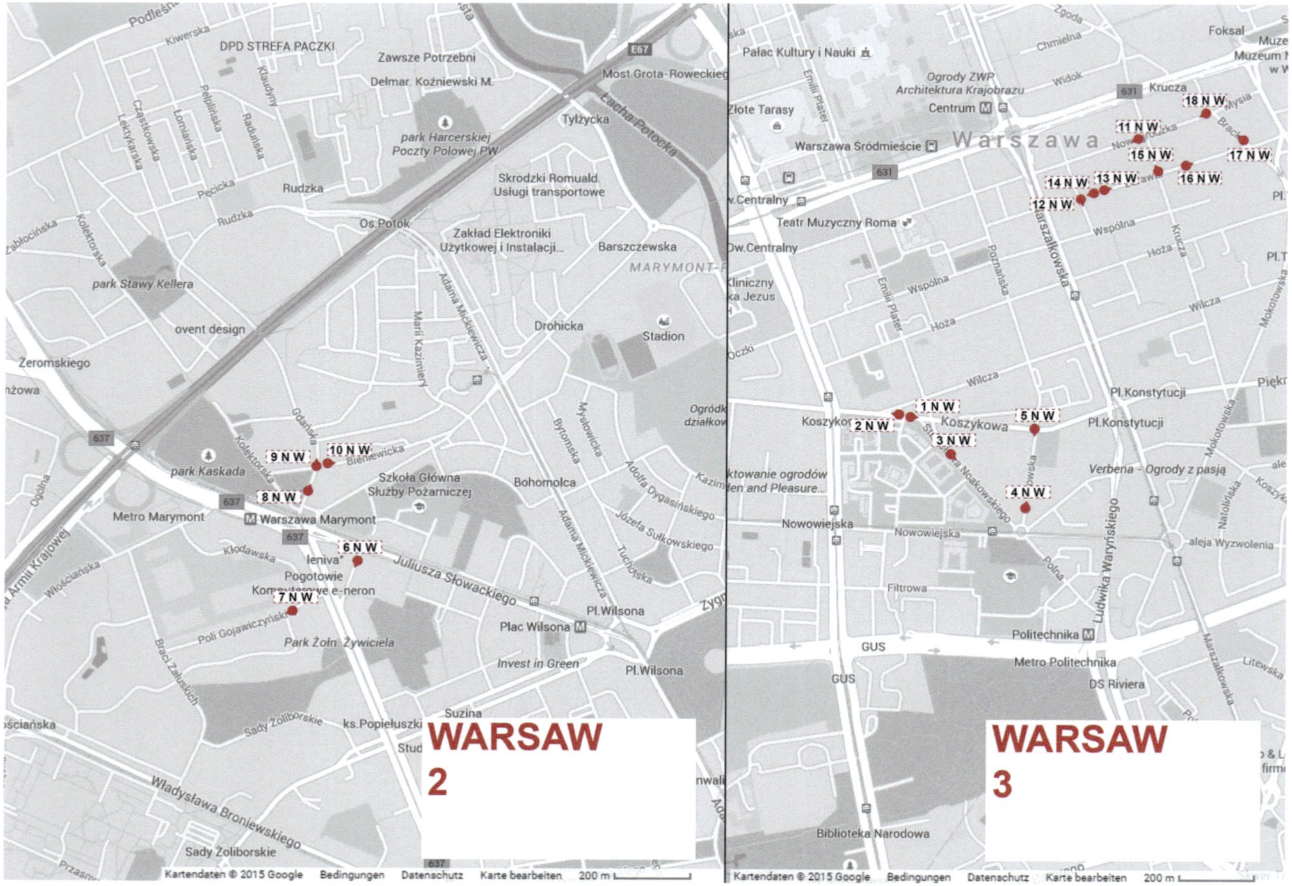

Fig. 4.78 Locations for photo sessions in Warsaw. (Source: Google Maps, edited by the author)

4.5.2 Impressions During the Site Inspection

The expressed expectation that Warsaw has very high levels of brightness after dark is confirmed (Fig. 4.79). This is due to the large number of pole-top luminaires (Fig. 4.80), the dense spacing of the luminaires (Figs. 4.79 and 4.80), the excessive lighting of the facades, shop windows/entrances (Figs. 4.79 and 4.80) and the brightly lit advertising spaces (Fig. 4.79). And clutter is in evidence in various places.

The mounting height of the luminaires is approximately 10–12 m, which is very high, and similar to Toronto. When viewed on site, the street space appears to be uniformly illuminated (Fig. 4.80).

The luminaires in the area do not produce any glare, but glare is caused by the illumination of the advertising surfaces. The advertising spaces are illuminated from the side and from top to bottom. The luminaires used here are poorly shielded, which leads to disturbing glare (Fig. 4.81).

The overall advertising space is perceived as very bright. This can be explained by the fact that these are usually light in colour and the paper the advertising is printed on has a high degree of reflection. In addition, the surfaces are illuminated using cool white light. The cool white light, as explained above, is often perceived as being brighter than warm white, or yellowish, light (see Sect. 3.1) due to its blue-green component. Moreover, it can

Fig. 4.79 Location for photo session 14 H W

Fig. 4.80 Location for photo session 3 H W

Fig. 4.81 Location for photo session 4 H W

Fig. 4.82 Location for photo session 13 H W

be assumed that the luminaires used to illuminate the advertising surfaces feature increased luminous flux so that they can achieve increased presence in the space.

The facade lighting (if present) is only provided from the outside. The facades are generally illuminated using uplights, whereby these are often incorrectly adjusted and a large part of the light is directed into the atmosphere. This light trespass contributes to light pollution (Fig. 4.82, the light is not cut off at the edge of the roof but radiates over it).

Extreme levels of brightness were observed at shop windows/entrances (Fig. 4.79).

No differences in the light distribution curve of the luminaire types under study could be determined on site.

The four types of luminaire are equipped with sodium vapour lamps. In the narrower streets there are some exceptions where the luminaires are equipped with metal halide lamps (e.g. in the cul-de-sacs at the Technical University or near Pałac Kultury i Nauki, Fig. 4.82).

Contrary to previous experience in cities, these sections appear darker than those lit by sodium vapour lamps.

The assumption that the W B luminaire is a further development of the W A luminaire is confirmed. The photometrics of the two luminaires appear to be identical.

In the 3–6 H W area, a new luminaire type W E (Fig. 4.83) was recorded during the site inspection: a metal halide luminaire with a cutoff housing. The urban space equipped with these luminaires features a lower level of brightness compared to other urban spaces. The

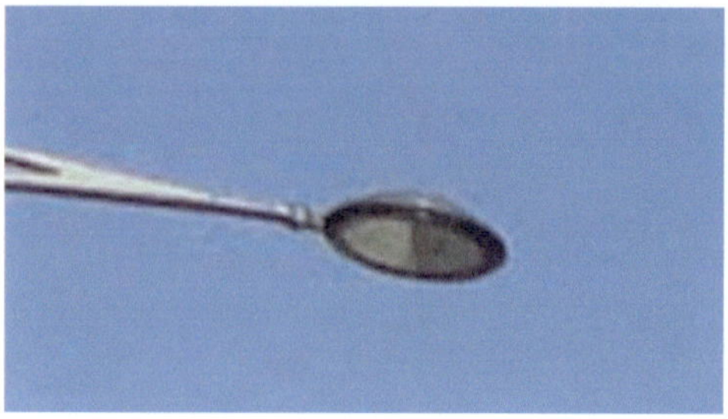

Fig. 4.83 Luminaire type W E

Fig. 4.84 Location for photo session 10 N W

cutoff housing is responsible for this. Cutoff luminaires direct more light onto the horizontal surfaces and less towards vertical surfaces. As explained in Sect. 4.2.3, this can lead to a different brightness perception of the space.

The side roads in Warsaw tend to be darker than the main roads. This is likely due to the fact that there is less illuminated advertising space in these roads. Furthermore, the facades are not actively illuminated. Illumination is provided almost exclusively via the street lighting (Fig. 4.84).

Similar to New York City, the main roads are also highly frequented well into the night (observations from 11.00 pm to 1.00 am). Lighting from motor vehicles will thus also contribute to illuminating the space. Above all, stationary traffic at red traffic lights increases the level of brightness in the urban space, given the multitude of stationary cars.

4.5.3 Evaluation of the Measurement Data

It should be mentioned at the outset that location point 14 H W for taking photos was not included in the assessment of average luminance in the urban space. As you can see from Table A.7 in the appendix, this point of measurement produced extreme values in terms of luminance. Compared to the other photos the values are disproportionately high and therefore do not correspond to the average values for the urban space in the City of Warsaw. Accordingly, the values were not included in the rating (similar to Times Square in New York City).

The luminance images show at first glance that a particularly high brightness level prevails here (Figs. 4.86 to 4.92). The average values in Fig. 4.85 support this. Compared to the values of the previously analysed metropolises, the values in this case are more than doubled. The forecast made in the preliminary analysis and the impressions on site are thus objectively substantiated by the measurement data. As previously assumed, the urban space is subject to extremely high luminance levels because of the many clutter scenarios. These can be seen in detail in Fig. 4.86. The surfaces with the highest luminance in the image are the streets and sidewalks. These receive light from facades, shop windows/entrances and street lighting. There is way too much street lighting. There are often several luminaires mounted on one pole and the spacing between the poles is too tight.

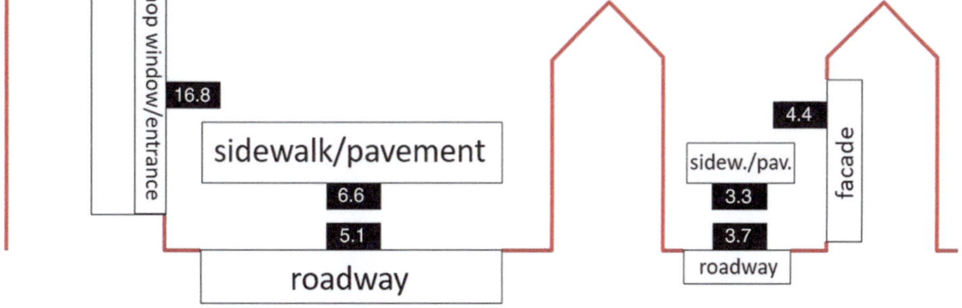

Fig. 4.85 Average luminance (cd/m^2) in urban areas, Warsaw

Fig. 4.86 Location for photo session 3 H W

The number of different light sources within a confined space gives rise to extreme levels of brightness in the urban realm. The luminous flux of the light sources is as high as that in the other cities in the comparative analysis (identifiable in the false colour images). This confirms that the high number of luminaires is the main cause of the "overlighting".

In general, it can be confirmed that the luminance images indicate that the lighting in the urban space is uniformly spread (e.g. Figs. 4.86 and 4.87). This is generated by the large number of luminaires, the dense spacing of the light poles, and the mounting heights of the luminaires. This leads to the light cones of the individual luminaires overlapping across a large area, creating the impression of uniform lighting.

Figure 4.88 shows that the advertising spaces are strongly illuminated and several times brighter (approx. 15 cd/m^2) than the already strongly illuminated urban space (approx. 7 cd/m^2). As previously speculated, both the surface quality of the advertising surfaces and the total luminous flux of the luminaires play a role here, leading to this result.

In addition, a luminance of 480 cd/m^2 of the advertising surface illumination was recorded from the location for the photo session. This value is almost 68 times higher than the surrounding area at approximately 7 cd/m^2. As a result, the advertising spaces are perceived as sources of disturbing glare (see Sect. 3.1).

One factor revealed by the measurement data is the excessive illumination of individual facades in the area. For example, the facade in Fig. 4.86 has a maximum value of 395 cd/m^2 and the facade in Fig. 4.89 even a maximum value of 538 cd/m^2. Especially in the sections with these types of building (buildings with illuminated facades), this impacts the overall

Fig. 4.87 Location for photo session 17 H W

Fig. 4.88 Location for photo session 4 H W

Fig. 4.89 Location for photo session 13 H W

brightness in the space considerably. This aspect was not perceived to the same extent on site.

The distribution curves of the individual luminaires could not be analysed during the evaluation. Due to the large number of luminaires and the dense spacing of the light poles, the various distribution curves overlap. It is therefore impossible to determine whether the different luminaire types differ from one another in their beam characteristics (Fig. 4.90).

Contrary to the experience gathered on site, the spaces with cut-off luminaires equipped with metal halide lamps tend to be brighter than the urban spaces with sodium vapour lighting (Fig. 4.87 sodium vapour luminaires and Fig. 4.91 metal halide luminaires). This may be because of other reflectance factors in the street sections or because the spaces appear to be more open, as shown in the photo session locations 7–10 H W. The distances between the facades play an important role with regard to reflections (see Sects. 4.2.3, 4.3.2 and 4.3.3).

The measured data reinforce the thesis that the side roads are on average darker than the main roads. The lighting schemes do not differ between the main and side streets. As a result, they have a similar basic level of brightness. As there are fewer shops and advertising spaces in the side streets, and the brightness level here is lower compared to the main streets (Fig. 4.92).

The highest luminance values recorded in Warsaw were in the shop window/entrance category. The average value here is approximately 16.8 cd/m^2. In addition, lighting clutter presents a further problem for the metropolis.

Fig. 4.90 Location for photo session 15 H W

Fig. 4.91 Location for photo session 7 H W

Fig. 4.92 Location for photo session 10 N W

4.6 Hanover

4.6.1 Preliminary Analysis, Expectations

Hanover has a population of approximately 500,000 people (cf. Emporis 2015b). The city represents a typical medium-sized European city. Similar to Warsaw, the streets and urban spaces developed more or less naturally and were not subject to a higher principle of order, such as a grid, as is the case in many North American cities.

The city has low building heights compared to other metropolises. The highest buildings, according to emporis.com, are the residential buildings Ihmeplatz 2 and Lister Tor (excluding television towers), which have a height of approximately 90 m (ibid.).

In the course of the preliminary research no satellite pictures of the city could be found. On the basis of the knowledge gained during the analyses of the previous cities, it was assumed that the lighting hotspot would be located in the centre of the city. Accordingly, the luminance measurements were carried out in the centre of the city (in the vicinity of the central station).

Compared to many other European cities in this order of magnitude, the core of the centre is basically a traffic-free zone with shopping facilities (Fig. 4.93 representative image).

Although there is no car traffic in this space, these pedestrian areas are defined as main streets given that they are highly frequented (pedestrian precinct from the main station to Osterstrasse). In addition, larger streets, which indicate a high traffic load, are also defined as main roads.

Fig. 4.93 Location for photo session 19 H H

These roads generally have two lanes in each direction. The roadway is approximately 3 m wide, with a pavement area with a width of around 4 m. No dimensions can be provided for the pedestrian areas, since they differ considerably.

The side roads in this case comprise one or two lanes with a width of 2.75–3 m. Here, too, there are roads in which road traffic is severely restricted. However, these streets consist mainly of residential buildings, which means that they are to be regarded as side streets. The width of the sidewalks cannot be specified here either. Here, too, there are considerable differences in the dimensions. In addition, in some of the side roads there is no clear separation of pavement and roadway.

A special feature that was noticed during the research is the high density of trees and the resulting positioning of the luminaires. These are located either between the treetops or above the roadway (Figs. 4.94 and 4.95).

There are six different types of luminaires in the area under study: luminaire type H A (Fig. 4.96), luminaire type H B (Fig. 4.97), luminaire type H C (Fig. 4.98), luminaire type H D (Fig. 4.99), luminaire type H E (Fig. 4.100), luminaire type H F (Fig. 4.101). All luminaire types are semi-cutoff luminaires which are presumably fitted with sodium or metal halide lamps.

Luminaire types H A and H B are the most common and are mainly present along the main roads. H A type luminaires are either mounted on poles or suspended above the road. H B type luminaires are mounted on poles, whereby at crossings there are three luminaires mounted on one pole.

Fig. 4.94 Location for photo session 1 H H, on site

Fig. 4.95 Location for photo session 15 H H, on site

Fig. 4.96 Luminaire type H A, on site

Fig. 4.97 Luminaire type H B, on site

Fig. 4.98 Luminaire type H C, on site

Luminaire type H C is only to be found in the side streets, generally at a mounting height of approximately 5 m.

H D type luminaires are located in the city centre and are installed both in the main traffic-free zones (pedestrian precincts) and in the main roads with high volumes of traffic.

The H E lighting columns and H F type luminaires are only installed in traffic-free zones. The lighting columns are only installed in the core of the centre, whereas H F luminaires can also be found outside the centre (Lister Mile).

The buildings in the main streets only contain retail and office space, especially in the city centre. The percentage of living space increases the further one moves away from the city centre.

The side roads in the area consist almost exclusively of residential streets.

Fig. 4.99 Luminaire type H D, on site

Fig. 4.100 Luminaire type H E, on site

The predominant facade materials in the city centre are light-coloured stone and glass. Outside the centre, however, plaster and brick facades predominate in places (Fig. 4.102).

Following the first impressions gained in the research, it was expected that many places would feature "patchy" lighting. This is basically due to the high number of trees. It can also be assumed that the combination of trees and light will give rise to some positive aspects (cf. Boston).

Politically speaking, it can be read in several newspapers and on the Internet that the SPD and the Greens are campaigning for a "darker" Hanover at night (cf. Nicolay 2015; Schinkel 2015). This leads to the conclusion that Hanover has a light pollution problem. As a result, bright urban spaces are expected.

Figure 4.103 shows the locations selected for the photo sessions in Hanover.

Fig. 4.101 Luminaire type H F, on site

Fig. 4.102 Facade surfaces, on site

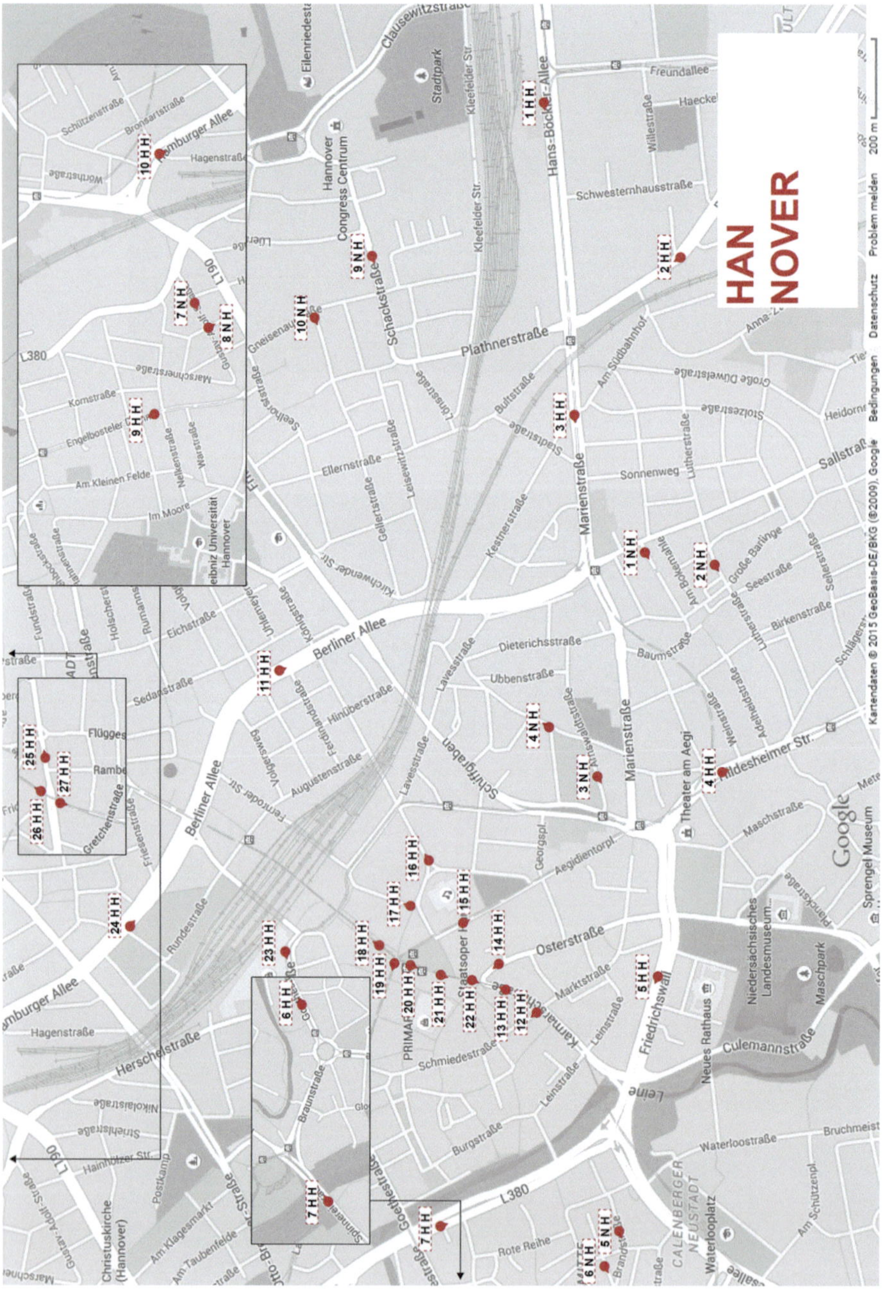

Fig. 4.103 Locations for photo sessions in Hanover. (Source: Google Maps, edited by the author)

4.6.2 Impressions During the Site Inspection

The assumption, derived from newspaper articles and Internet sources, that Hanover is an especially bright city is not confirmed on site. In districts featuring avenues the opposite is rather the case. Despite the low level of brightness, the urban spaces are adequately illuminated. Orientation within the space and a sense of security are guaranteed. There are no areas where the level of brightness is too low.

Hanover was perceived all in all as the "darkest" city in the analysis.

The main roads with and without vehicle traffic have a noticeably higher level of brightness than the side roads (Fig. 4.104 main road, Fig. 4.105 side road). The street luminaires used in the side streets (luminaire type H C) have a lower lumen package. From this it can be concluded that the lighting in the main and side roads was planned differently. This presumption can be justified by referring to DIN 13201. According to the standards, the level of brightness is adjusted to align with the extent to which the street is used or frequented. The more traffic there is expected, the brighter the lighting will be. Since the side roads in the area are mostly residential streets, the lighting is kept correspondingly low.

In general, the illumination of the city is perceived as very homogeneous. As already assumed in the preliminary analysis, the sections with a dense tree population are in some cases heavily shaded by the foliage. The position of the luminaire is responsible for the degree of shadow generated. If the mounting height is below the treetops, branches or

Fig. 4.104 Location for photo session 23 H H

Fig. 4.105 Location for photo session 1 N H

leaves are not in the way (Fig. 4.106). If the mounting height is at the same level as the treetops, however, this will naturally cast shadows as a result (Figs. 4.94 and 4.95).

Luminaire types H C and H B are sodium vapour luminaires. Luminaire type H A is used outside pedestrian precincts equipped with a sodium vapour lamp, and in pedestrian precincts equipped with a metal hailde lamp.

Luminaire types H D and H F, which are installed exclusively in the pedestrian precincts, are also equipped with metal halide lamps. Thus the city centre is illuminated exclusively using metal halide lamps.

Luminaire type H F is also a metal halide luminaire. In the course of the site inspection, it was found that this luminaire type had the lowest mounting height of all luminaires under study. This is advantageous for the rows of trees (Fig. 4.106, no shadows cast by the foliage).

Differences in brightness, between urban spaces with sodium vapour luminaires on the one hand and metal halide luminaires on the other, cannot be detected visually. Excessive illumination of the facades was not in evidence.

Without exception, the luminaires in the area are not perceived as generating glare.

The shop windows/entrances are not excessively illuminated. The shop window lighting is often switched off at night. This kind of lighting therefore has little impact on the brightness of the urban space (Fig. 4.106).

Fig. 4.106 Location for photo session 27 H H

4.6.3 Evaluation of the Measurement Data

The values in Fig. 4.107 confirm the impression on site. The measurements have shown that Hanover is the least "bright" of all the cities under study. The fact that the city as a whole has low values in all valuation categories, which hardly differ from each other, allows the eye to adapt to the level of brightness. Despite the low brightness, the room is perceived as sufficiently bright thanks to the adaptation of the eye (see Sect. 3.1).

The uniformity of the illumination can also be confirmed by the luminance images. However, it must be acknowledged that this uniformity of lighting is to a lesser degree than in Warsaw. In Hanover there are smaller lighting hotspots beneath the luminaires (Fig. 4.108).

The same can also be seen in the side roads. The lower level of brightness in the side roads compared to the main roads is confirmed throughout the area (Fig. 4.109).

As might be expected, most roads with trees feature less uniformity due to the shadows generated by the treetops (Fig. 4.110). Where the luminaires are installed at lower mounting heights, however, the trees have a positive effect on the lighting in the urban space, as is the case in Boston (see Sect. 4.3.3). In the case of the H F luminaire, the uniformity of the lighting in the space is only minimally influenced by the low mounting height (Fig. 4.111). The values indicate that roads with trees tend to have a lower average luminance than roads without trees. This is due to shading, which reduces the average value.

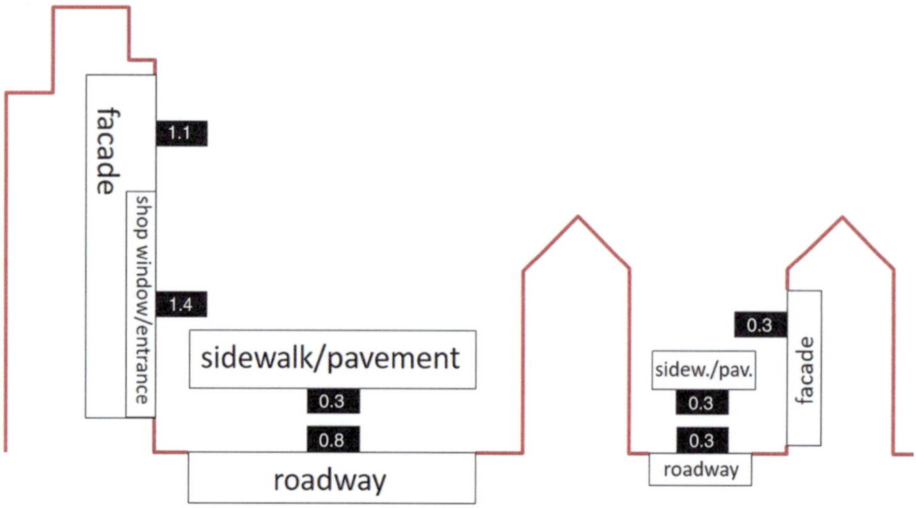

Fig. 4.107 Average luminance levels (cd/m^2) in the urban realm, Hanover

Fig. 4.108 Location for photo session 23 H H

Fig. 4.109 Location for photo session 6 N H

Fig. 4.110 Location for photo session 1 H H

Fig. 4.111 Location for photo session 27 H H

Figure 4.112 shows a clutter scenario. This results in increased overall emissions from the shop windows and the H D luminaire. At approx. 13.5 cd/m^2, this value is well above the city average.

Figure 4.112 shows that the luminaire type H E only functions as a decorative luminaire. The space is not influenced by the light emitted by the luminaire. This can be seen from the fact that at the points where the luminaires are defective, the same light patterns are produced on the ground and walls as at comparable points where the luminaires are intact.

A further clutter scenario is in evidence at the 13 H H photo session location (Fig. 4.113). Here it can be seen that there are two H D luminaires mounted on one pole. In conjunction with the densely spaced poles and the narrow cross section of the street, this gives rise to increased brightness levels.

On the basis of the photographs it can generally be said that the H D luminaire produces more light directly below the luminaire compared with other luminaire types (Fig. 4.108). This has a negative effect on the uniformity of the lighting in the space. It should also be noted that the luminaires are better shielded than the other types of luminaires. This is also shown by the fact that the luminaire head features less luminance compared to the other luminaire types (the "glow" or "ball of light" around the luminaire head is smaller).

The greatest uniformity in terms of light distribution was found in road sections where H A luminaires were installed. The luminaires are mounted at a considerable height in these areas, which is advantageous in terms of uniformity.

Fig. 4.112 Location for photo session 18 H H

In addition, the glass diffuser is bowl-shaped. In conjunction with an appropriate reflector, this generates very uniform light emission (Fig. 4.110).

The photographs of the H F luminaire show that this luminaire also delivers good uniform lighting in the urban realm (Fig. 4.111). The luminaire radiates its light uniformly in all horizontal directions. The canopy on top of the housing prevents any spill light into the sky.

Figure 4.113 shows the highest measured value in the category shop windows/entrances: approximately 14 cd/m^2. Similar to New York City, this area is illuminated above average. It is rare in Hanover for shop windows/entrances to be illuminated with such intensity. Figure 4.114 is an excellent example of the lighting of shop windows in the city. As can be seen in the picture on the left, there are a large number of shop windows and entrance passages whose lighting is switched off at night. Although the shop window/entrance category also features the highest luminance values, it can basically be ignored. The percentage of this category is too small in the city as a whole. Clutter plays a far more significant role, as can be seen in Figs. 4.112 and 4.113.

Fig. 4.113 Location for photo session 13 H H

Fig. 4.114 Location for photo session 20 H H

4.7 Final Comparison of All Metropolises and the Conclusions Drawn from It

Firstly, it should be emphasized that each city has different problems regarding light emission and pollution. At this point of the study, it is already evident that there is no all-encompassing measure that can be applied to ensure a reduction in light pollution at all locations. Each of the metropolises must develop its own solutions.

The city analysis shows various urban scenarios that have a positive or negative impact on light pollution. In this section these will be explained by means of summarised comparisons of the metropolises. The findings serve as a basis for the development of individual cities' own solutions in later chapters.

The studies of the different cities show that the brightness of the road surfaces in the main roads does not differ (with the exception of Warsaw). They have values of between approximately 1 and 1.5 cd/m^2, whereby these do not represent excessive values (see Sect. 4.2.3, reference to DIN 13201).

Despite the similar results, there are strong differences in the perception of the entire urban space. Boston, for example, is perceived as being considerably darker than New York City, which the measurements refute, however. In the analysis, various factors have emerged that influence and explain the perception of brightness.

Basically, it can be stated that the perception of brightness is not a fixed unit of measure, but a response governed by human senses. Thus, the perceived scenario can differ from person to person.

A generally important factor that has an influence on perception is the material the facades are made of. In Boston, for example, there are predominantly dark materials compared with New York City. As a result, the urban space as a whole appears to be darker. In addition, there are only a few surfaces with highly reflective properties (e.g. glass). This gives rise to very few reflections, which would cause an increase in brightness in the space.

The height of the buildings also plays an important role. The buildings in New York City are far higher than those in Boston. If an area consists mainly of tall buildings, only a small part of the dark sky can be seen from a human perspective. Mainly facade surfaces will be visible in the field of vision. If the buildings are lower in height, the percentage of the dark sky within the field of vision increases. The higher percentage of dark sky there is in the field of vision, the darker the urban space may appear to be. That is to say, the urban realm is perceived as darker than the measured values indicate. This shows that vertical surfaces are more important for the perception of space than horizontal surfaces.

Another example that further underscores this can be seen in Boston (Fig. 4.44). The horizontal surfaces are illuminated relatively strongly, but the full cutoff luminaires do not allow illumination of the facades. As a result, the urban space appears to be dark.

The analysis also shows that streets lit by luminaires equipped with metal halide lamps and LEDs are perceived to be brighter than areas illuminated using sodium vapour lamps (Boston, Toronto). This can be explained by the spectra of the respective lamp technologies. At night, the human eye reacts more sensitively to blue-green components in the light, making these light sources appear brighter than light sources with a high yellow-red component, even given the same luminous flux (see Sect. 3.1). Thus, the specific light spectra also play a significant role in the perception of brightness.

An important factor in terms of brightness is the average overall level of brightness that prevails in a city. If this is generally low, the eye adapts to that level. Thus, just as much can be perceived as in urban spaces with higher levels of brightness. This is the result of the comparative analysis between Hanover and Warsaw.

The studies undertaken in Toronto and Boston show the influence of road cross-sections on brightness. The side roads in the two cities are brighter than the main roads, whereby the side roads have the same luminaires as the main roads. The fact that the facades are closer together, however, means that considerably more light is reflected, which makes the space much brighter.

In New York City and Toronto it can be observed that tall buildings have great potential for generating light pollution due to their expansive facade surfaces and backlit windows. In the analyses the window surfaces that emit light outside have a luminance of approximately 5 cd/m^2.

The reflective facades that predominate in Toronto and New York City also have great potential for generating light pollution, as the analyses have shown. These can add substantially to the level of brightness in the urban realm. This means that there must be luminous objects in the urban space that can be reflected. In this respect, New York City is affected more severely than Toronto. In contrast to Toronto, New York City has a large number of buildings that are illuminated, which means that they can potentially be reflected by the reflective facades (Fig. 4.21).

With regard to the illumination of the facades, the studies undertaken in Warsaw and New York City show that this adds considerably to the amount of light in the urban space (Figs. 4.23, 4.88 and 4.89). Lit vertical surfaces (including advertising surfaces) have the highest luminance values in that category in the cities.

With regard to the uniformity of the illumination of the urban spaces, it should be noted that the mounting height and positioning of the luminaires play a decisive role. In the urban spaces featuring uniform lighting the mounting heights are generally high (Toronto, Warsaw, Hanover). This enables the light to be spread more effectively within the space. Wider light distribution curves can be applied, which support the uniform lighting effect. In Warsaw, uniform lighting can be experienced at an increased level: the excessive number of luminaires and the extremely dense spacing of the poles gives rise to a large number of beam spread overlaps. The result is ultrabright uniform lighting in the urban space.

Low mounting heights, on the other hand, lead to spot-like beam spreads. The space is perceived as being less uniformly illuminated (Figs. 4.66 and 4.67).

The optimum position for the light source is directly above the surface to be illuminated. In Hanover, a luminaire is positioned centrally above the carriageway (Fig. 4.110). The road space is therefore uniformly lit (with the exception of the shadows created by the foliage). In contrast, in Fig. 4.44 you can see a luminaire located at the side of the roadway.

As a result, the road is illuminated less evenly (the effect is also enhanced by the fact that this is a full cutoff luminaire).

In Hanover and Boston, it can be seen that the correct positioning of luminaires in combination with trees produces beneficial effects. The cities have tree-lined avenues, whereby the luminaires are mostly to be found below the treetops. This constellation ensures the space beneath the treetops is well lit. Very little light escapes into the atmosphere, however (Figs. 4.47 and 4.111).

It should be emphasized that every city under analysis was observed as being negatively influenced by clutter. The street lights in New York City are strictly spaced. Here clutter is not caused by luminaires positioned too closely together, as is predominantly the case in other cities. In New York City, the clutter occurs through the combination of the luminaires and shop windows/entrances being so close to each other (Fig. 4.20). This scenario can also be found in the other cities.

Light pollution as a result of light trespass was only detected in the analysis in Warsaw (Fig. 4.82). This aspect is therefore less relevant for the metropolises under study.

The photo taken at location 4 N B (Fig. 4.48) is of special interest. As mentioned in Sect. 4.3.3, this is a side road that does not have any street lights. Sufficient brightness is generated by the shop windows and the residual brightness from the surrounding main roads. This raises the question as to whether parts of the side roads could be planned without street lights. Especially cities with basically high light levels, such as Warsaw, could be considered.

In Warsaw and New York City, it was observed that stationary traffic also contributed to the level of brightness in the urban space. The extent to which the space became brighter could not be detected by the luminance camera. However, the aspect of such dynamic effects should not go unnoticed and should be taken into account when later developing solutions.

The initial speculation that cultural differences may also influence lighting (see Sect. 4.1) cannot be confirmed by the analysis. The values recorded in the North American cities are similar and can therefore be summarized. These differ markedly from the values in Germany and Poland. Warsaw has by far the highest lighting levels, and Hanover the lowest. The North American cities are in between. These differences will, however, be due to different regulations in the individual metropolises or countries, and not to different cultures or mentalities.

To sum up, it can be said that the on-site photos were able to show the different light pollution problems of the individual cities. Furthermore, it was possible to identify urban situations which have a positive or negative effect on light pollution and thus make a major contribution to the development of our own solutions. It is also possible to compare the cities in each category. However, no statement can be made about how much light actually reaches the sky from the individual categories/surfaces. The images predominantly reflect the light pollution potential of the individual urban spaces, and bear no reference to the size of the area. For example, one category may have particularly high values, but in the overall context of light pollution it may have only minor relevance. A good example of this is

Table 4.1 Brief summary of the conclusions

Urban components/ features	Impact
Facade surface (material properties) Building density Building height Light spectrum of luminaire Prevailing average brightness	Different urban spaces appear to be brighter or darker although measured values are the same
Cross-section of road	Perceived brightness under the same lighting conditions: Narrow = brighter, wide = darker
Skyscraper/highrise facades	Extensive facades with illuminated windows, facade illumination and reflective facade surfaces have great light pollution potential
Illuminated vertical surfaces	Above average light emission/light pollution
Mounting height Position of luminaire Luminaire spacing Light distribution curve	Influences uniformity of the lighting Influences the development of clutter and light trespass
Luminaire-tree combination	Good level of brightness with little light emission into the sky
Shop window/light combination	Potential for clutter
Brightly lit major roads	Enables street lights to be left out of side roads
(high volume of traffic)	(raises the level of brightness in the urban realm)

Hanover (see Sect. 4.6.3). The shop windows/entrances category has the highest values, but the surface area in this category is so small that it will have little relevance in the overall context of urban light pollution. In order to be able to make statements about the degree of light emitted in the individual categories, simulations of urban models with light scenes have been created in the following chapter.

The resulting conclusions from this chapter are set out in Table 4.1 below.

Bibliography

Cinzano, P./F. Falchi/C. D. Elvidge, “The first World Atlas of the artificial night sky brightness”, in: *Monthly Notices of the Royal Astronomical Society* Vol. 328, H. 3 (2001), pp. 689–707.

Cinzano, Pierantonio (Ed.), *Light pollution and the protection of the night environment. Proceedings of the conference Light pollution and the protection of the night environment Venice: let's save the night, held in Venice on May 3rd, 2002*. Thiene 2002.

City of Boston, LED Street Lighting, 31.07.2015a, (accessed 10.01.2017). http://www.cityofboston.gov/publicworks/lighting/led.asp

City of Boston, Lights out Boston, 31.07.2015b, (accessed 10.01.2017). http://www.cityofboston.gov/eeos/conservation/lightsoutboston.asp

Emporis, Boston Buildings, 31.07.2015a, (accessed 10. 01.2017). http://www.emporis.com/city/101045/boston-ma-usa

Emporis, Hanover, 06.08.2015b, (accessed: 10. 01.2017). http://www.emporis.com/city/100570/hanover-germany

Emporis, Toronto Buildings, 02.08.2015c, (accessed 10. 01.2017). http://www.emporis.com/city/100993/toronto-canada

Emporis, Warsaw Buildings, 04.08.2015d, (accessed 10. 01.2017). http://www.emporis.com/city/100803/warsaw-poland

FLAP CANADA, Hierarchy of Theats, 02.08.2015, (accessed: 10. 01.2017). http://www.flap.org/hierarchy-of-threats.php

Hiscocks, Peter, *Summary of Lighting Regulations City of Toronto,* 2011 Toronto 2011.

Marcellino, C., Governor signs Marcellino Bill to curtail Light Pollution from state Buildings, 18.12.2014, (accessed 10. 01.2017). https://www.nysenate.gov/newsroom/in-the-news/carl-l-marcellino/governor-signs-marcellino-bill-curtail-light-pollution-state

Nicolay, M., Zuviel Lichtverschmutzung, Rot-Grün will Stadt abdunkeln, 06.08.2015, (accessed: 10. 01.2017). http://www.bild.de/regional/hannover/dunkel/gruene-wollen-lichtverschmutzung-mindern-27498106.bild.html

Schinkel, A., Hannover ist Rot-Grün zu hell. SPD und Grünen ist es in der Stadt zu hell – sie fordern einen Vorstoβ gegen "Lichtverschmutzung", 06.08.2015, (accessed: 10. 01.2017). http://www.haz.de/Hannover/Aus-der-Stadt/Uebersicht/Hannover-ist-Rot-Gruen-zu-hell

Selene NY, Legislation, 29.07.2015, (accessed 10.01.2017). http://selene-ny.org/legislation/

United States Census Bureau, New York City Population, 28.07.2015, (accessed 10. 01.2017). http://factfinder.census.gov/faces/nav/jsf/pages/community_facts.xhtml

5 Simulation of Urban Scenarios/Modules and Spaces

5.1 Implementation of the Simulations

The simulations implemented serve to test and further develop the results gained through the studies described in Chap. 4. The simulations enable statements to be made about how much light enters the atmosphere from the respective categories.

Contrary to what was initially planned, a series of urban modules will be simulated and not a metropolis as a whole. This approach has the decisive advantage that the urban modules can be assembled in different constellations. Thus it is possible to reconstruct different cities using the selected urban modules. This will be demonstrated in Chap. 6. The simulations were carried out using Dialux 4.12 software.

A total of four urban districts have been simulated as 3D models:

- USA Main Road
- USA Side Road
- EU Main Road
- EU Side Road

The USA models have a distinct skyline. For example, they represent metropolises such as New York City or Toronto. The EU models consist of buildings that are overall lower in height. The structure is comparable to that of Warsaw, Hanover, but also Boston.

Four scenarios have been simulated for each model/urban module:

- Only street lights
- Only shop window lighting
- Only facade windows
- Combination of street lighting, shop window lighting and facade windows

E. E. Goronczy, *Light Pollution in Metropolises*,
https://doi.org/10.1007/978-3-658-29723-7_5

The research undertaken in Chap. 4 has shown that Boston and New York City are gradually equipping the streets with full cutoff lights (cf. City of Boston 2015; Selene NY 2015). These are intended to reduce light pollution. Also the associations "Vereinigung der Sternenfreunde e. V. Fachgruppe Dark Sky" and "The International Dark-Sky Association" recommend these luminaires, they are referred to as "Dark Sky friendly" (see Vereinigung der Sternenfreunde e. V. Fachgruppe Dark Sky 2015; The International Dark-Sky Association 2015).

As a result, all four scenarios were simulated with full cutoff and semi-cutoff luminaires.

It can be assumed that the relevance of shielding street luminaires decreases the taller the buildings are. The taller the buildings in the urban realm, the larger the vertical absorption surface. Much of the light produced by the street lights will therefore be absorbed into the urban space before it reaches the atmosphere. Also, the percentage of street luminaires in urban areas with tall buildings is very low compared to the percentage of backlit windows in facades. It can therefore be assumed that the majority of the total light emitted towards the atmosphere is generated by the lit windows in the facades.

The results of the simulations will show whether it is important for cities with tall buildings to invest in full cutoff luminaires.

In addition, three models were produced to clarify the influence of tree populations on light emission.

The simulation models (urban modules) and scenarios are explained briefly below. Only the relevant data and values are explained. Detailed information on the models, scenarios and results can be found online under "Urban Light Simulation Models, Emlyn E. Goronczy".

5.2 Structure of the Models

The models feature the material properties listed in Table 5.1.

The material properties are the preset default values that the Dialux program assigns to the individual materials.

Each model/urban module has a road with a fixed length of 200 m.

The models of the Main Roads have more lanes than the models of the Side Roads. A main road consists of: two pedestrian zones (each 9 m wide, including some space in front of the buildings), two lanes for use as parking bays or cycle paths (each 3 m wide), and four lanes (each 3 m wide) (Fig. 5.3). The side roads have two pedestrian zones (each 3 m wide), two lanes for use as parking bays or cycle paths (each 3 m wide) and two lanes (each 3 m wide) (Fig. 5.4). The cross section of the main roads is therefore wider. The measurements have been taken from the observations made in Chap. 4.

There are two blocks of houses on each side of the street (Fig. 5.1). For the simulation, the width of the housing blocks were defined as 85 m and the distance between them as 35 m (Fig. 5.3).

Table 5.1 Material properties

Object	Rho (reflection) (%)	Transparency (%)	Roughness (%)
Carriageway (roadway)	10	0	0
Bicycle lane/parking bay	15	0	0
Pavement/sidewalk	30	0	0
Facade	20	0	0
Tree trunk	19	0	0
Treetop	11	0	0

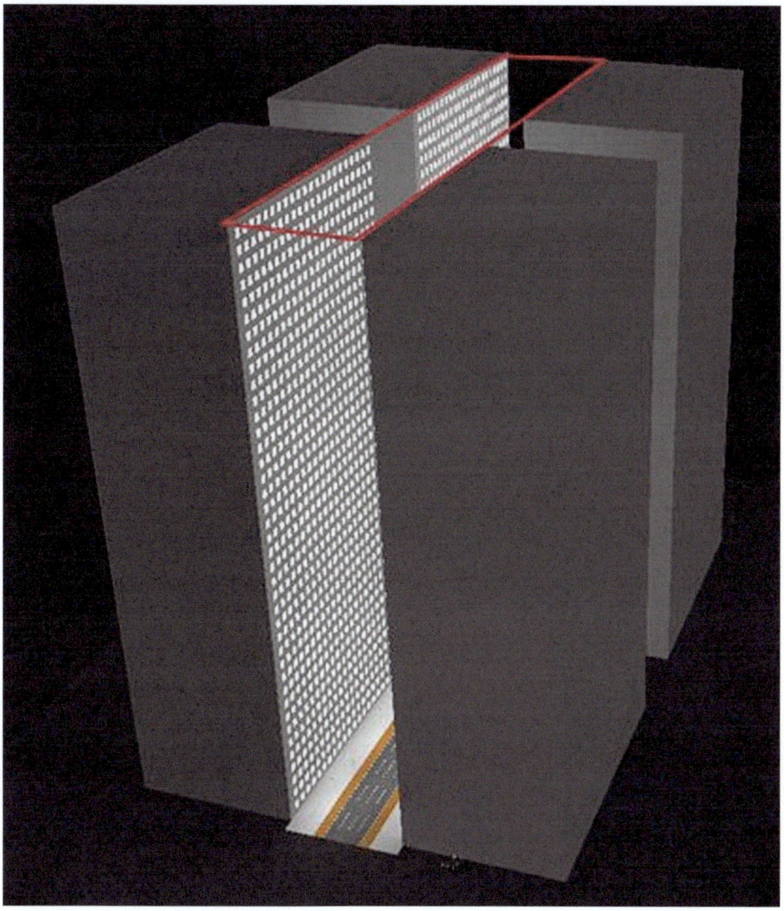

Fig. 5.1 Perspective of a main road in the USA

The building heights in the USA models are 180 m. This is the average height that could be determined via the website emporis.com. Emporis provides a list of the highest buildings in each city (Emporis 2015a, b, c, d, e, and f). For New York City and Toronto,

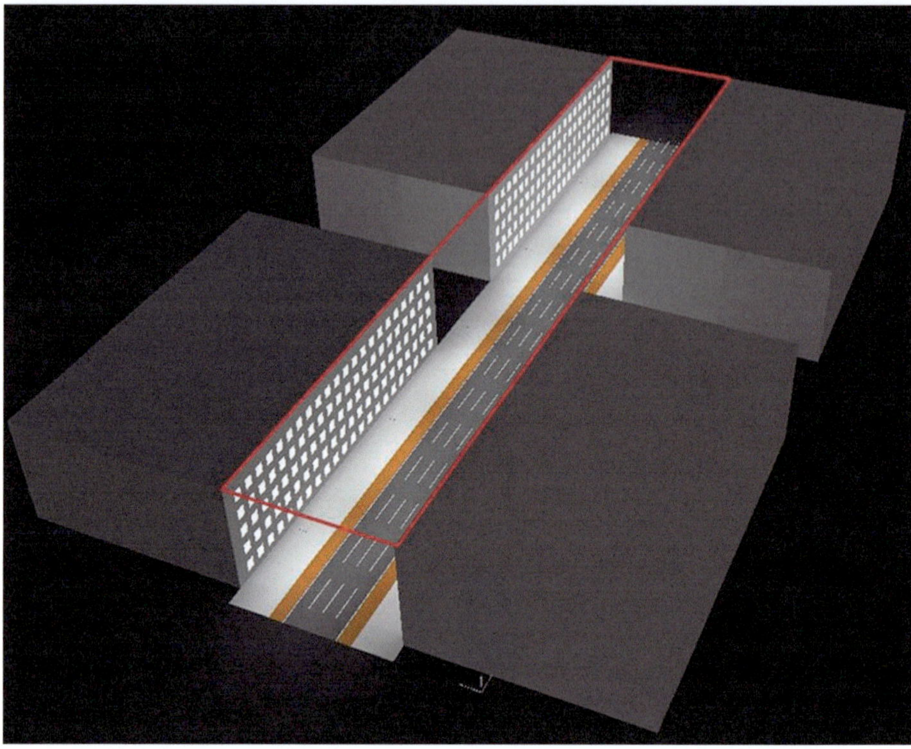

Fig. 5.2 Perspective of a main road in the EU

the ten highest, ten lowest and ten buildings in the middle of the list (in the skyscraper category) were included and the average calculated (see Appendix, Table A.3).

On the basis of the Emporis website, it could also be established that tower blocks with a height of 180 m have approximately 44 floors. This value was used for the simulation.

The urban modules of the EU models have a height of 25 m and five floors each (Fig. 5.2). These values were determined during the site inspection described in Chap. 4.

In all models, the windows in the facades are simulated using light-emitting rectangles. The luminance images in Chap. 4 show that illuminated windows have an average luminance of approximately 5 cd/m^2. Statements as to how many windows are illuminated on average at night cannot be made. After consultation with the lighting design team from Studio DL Warsaw, it can be assumed that the individual windows have a luminous flux of 200 lm. This is a realistic value, based on the observations made on site.

For the simulation model, the window dimensions have been set at 2 × 2.4 m on average (Fig. 5.3, enlarged view).

In each model, the two lowest rows of windows represent the shop windows. The findings in Chap. 4 show that the shop windows are on average two to three times brighter

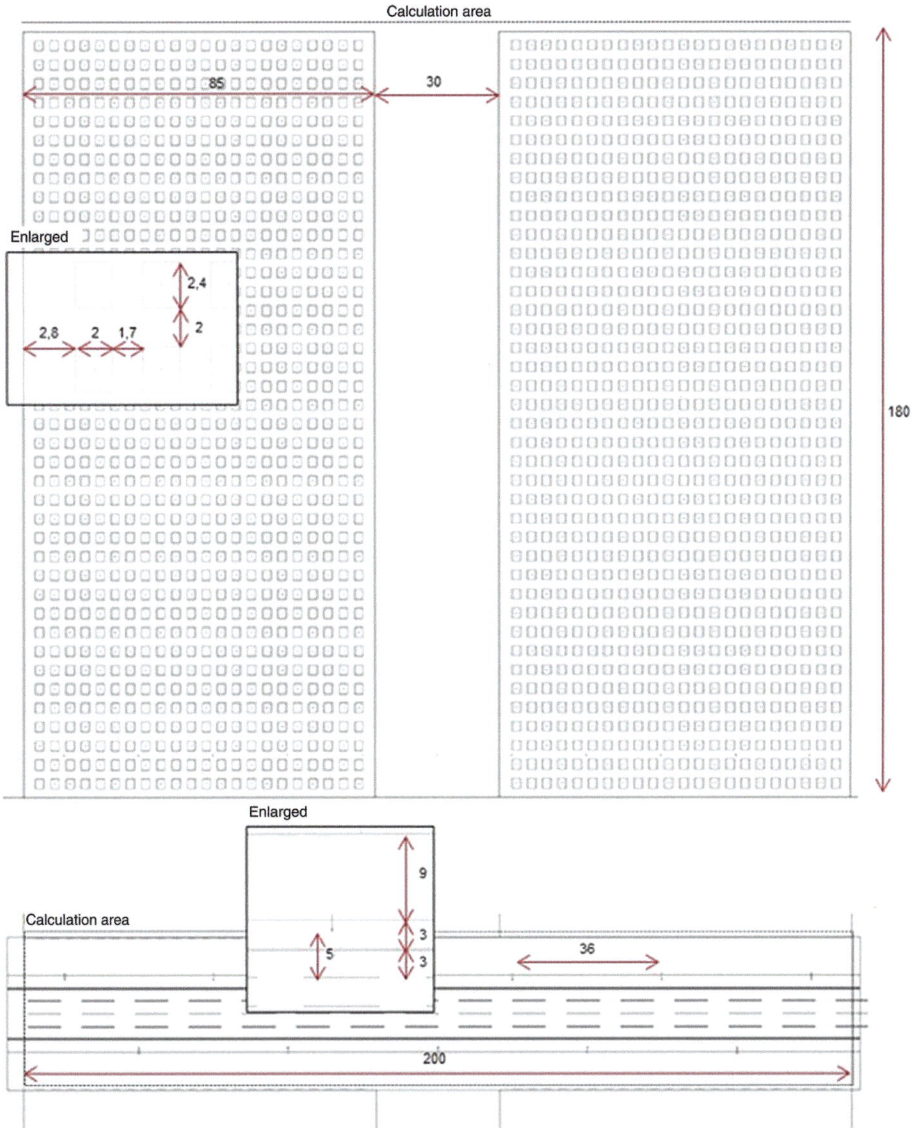

Fig. 5.3 Scaleless view and top view of the model of a USA Main Road, full cutoff, dimensions in metres

than "normal" windows. The luminous flux of the lower rows of windows has therefore been set at 460 lm per window.

In the simulations, the "7728" luminaire from Bega was chosen as an example of full cutoff luminaires. The luminaires are installed on both sides of the street: one side of the

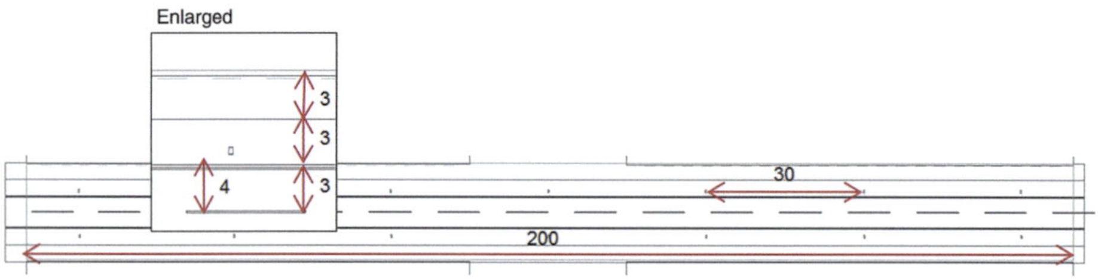

Fig. 5.4 Scaleless top view of the model of a USA Side Road, semi-cutoff, dimensions in metres

street has six luminaires and the other five. The distance between the luminaires is 36 m (Fig. 5.3, top view).

In the simulations, the Philips "Malaga SGS101-K 1xSON-I" luminaire was used as an example of semi-cutoff luminaires. The beam spread of the luminaires was increased by 15° in the direction of the street space in order to ensure higher illumination of the street space. In these simulations, eight luminaires (Fig. 5.5) were positioned on each side of the main roads and seven luminaires (Fig. 5.4) on each side of the side roads. The distance between the semi-cutoff luminaires is 25 m on the main roads and 30 m on the side roads.

The full cut-off luminaires on the main and side roads are positioned at a distance of 5 m from the outer edge of the roadway to the centre of the luminaire (Fig. 5.3).

In the case of semi-cut-off luminaires, the distance on the main roads is 3 m (Fig. 5.5) and on the side roads 4 m (Fig. 5.4).

All luminaires have a mounting height of 10 m. The different numbers and spacing of the luminaires is a result of the different light distribution curves.

The luminous flux of the street luminaires was modified so that a luminance of approximately 1 cd/m^2 is generated on the road surface. This is the average value for the luminance of roads as reported in Chap. 4. In addition, this value is also a planning value quoted in DIN 13201.

The calculation areas of the individual models align with the dimensions of the roads and are placed over the models, like a "lid" (Fig. 5.1, red frame, Fig. 5.2, red frame, Fig. 5.3). This makes it possible to measure the light emission towards the atmosphere.

The light-specific effects of tree-light combinations are tested using further simulation models.

Three different models were constructed, based on the "EU Side Road" model. The models consist of street lights aligned centrally above the roadway. The luminaires are Philips "Malaga SGS101-K 1xSON-I".

The first model (street without trees) contains only the luminaires in the street space, each 20 m apart (Fig. 5.6). In the second model (street with trees) trees were inserted, in close proximity to the lights (Fig. 5.7). In the third model (road with a larger number of trees) the trees are more densely spaced (Fig. 5.8). The mounting height in the models is 5 m. The trees have a height of 10 m (Fig. 5.9).

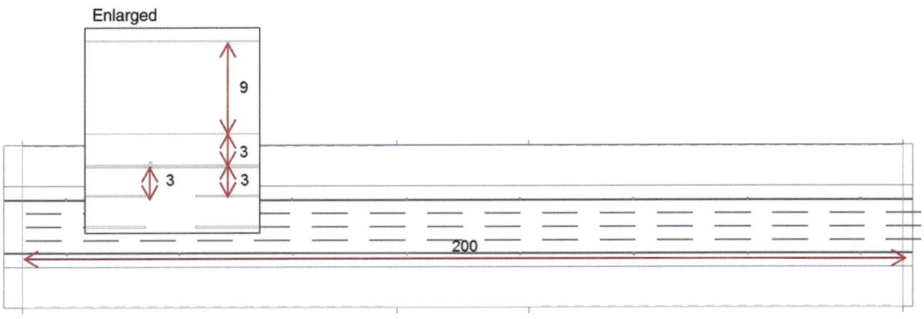

Fig. 5.5 Scaleless plan view of the model of a USA Main Road, semi-cutoff, dimensions in metres

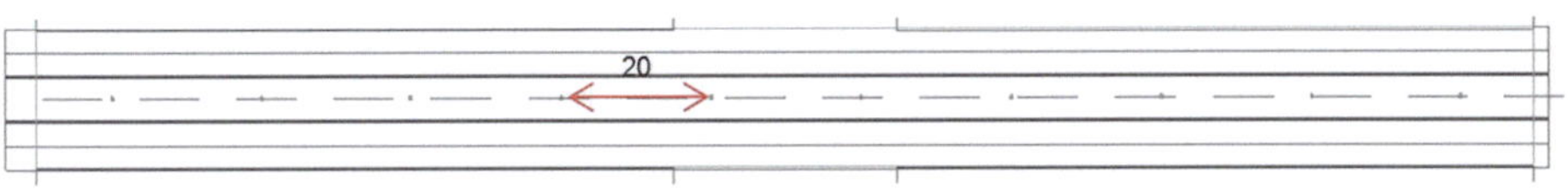

Fig. 5.6 Scaleless top view, model: without trees, dimensions in metres

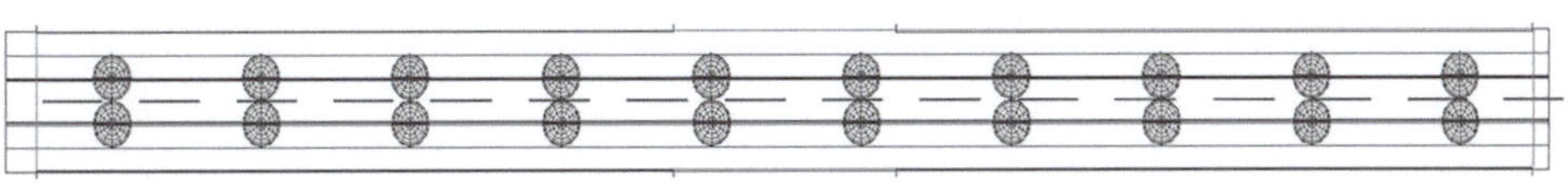

Fig. 5.7 Scaleless top view, model: with trees

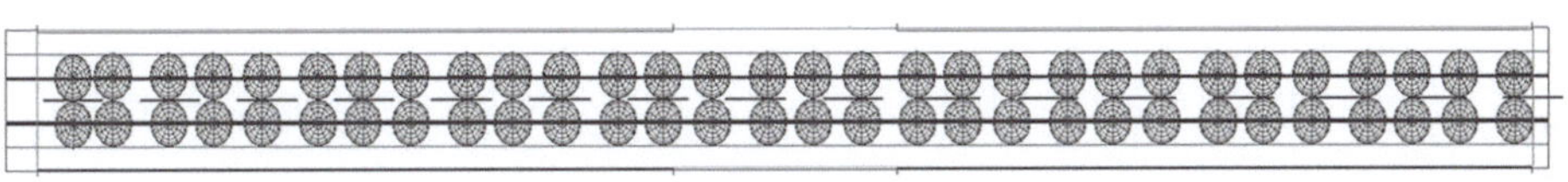

Fig. 5.8 Scaleless top view, model: larger number of trees

Two calculation areas were inserted into the simulation. The first one was placed 1 m above the road. The second one is positioned above the model in a "lid-like" fashion, as mentioned above (Fig. 5.9). This makes it possible to evaluate how much light there is in the space between the road and the treetops and how much light reaches, or is prevented from reaching, the atmosphere.

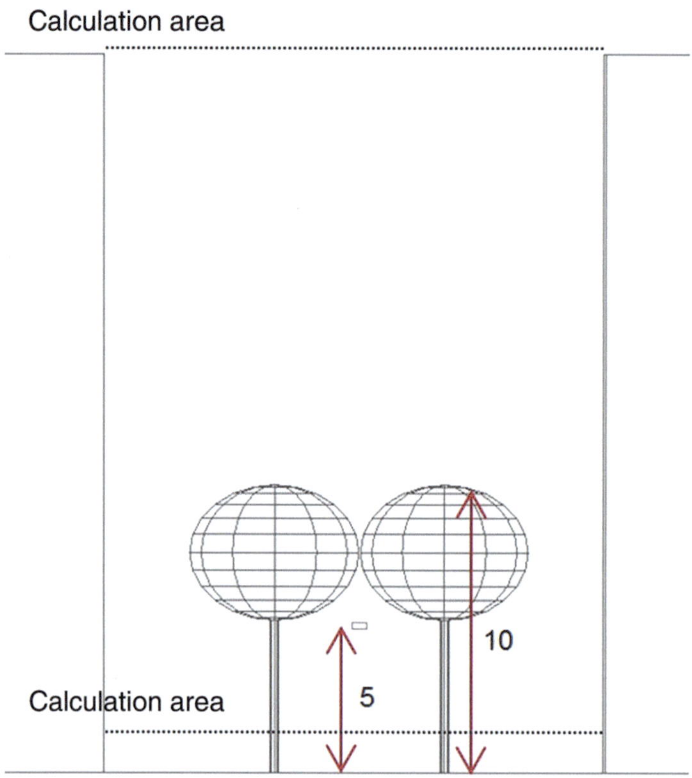

Fig. 5.9 Scaleless sectional drawing, model: with trees

5.3 Results of the Simulations of the Tree-Luminaire Combinations

The results of the simulations (Table 5.2) show that the tree-luminaire combinations have a positive effect in terms of light emission. When the luminaires are optimally positioned, the top of the tree absorbs most of the light. This prevents the light from entering the atmosphere.

Comparing the values of the "street without trees" model with those of the "street with larger number of trees" model, it can be seen that at almost the same illuminance in the streets in the "street with larger number of trees " model, far less light is emitted into the atmosphere. The overall reduction in light emissions is around 63%.

It should also not go unmentioned that merely by placing two trees next to each luminaire reduces light emissions by around 25% (the "street with trees" model).

Table 5.2 Simulation results of the tree-luminaire combination

Simulation model	Average illuminance into the sky (lx) (Calculation area over the model)	Average illumination 1 m above the road (lx) (Calculation area 1 m above roadway)
Street without trees	0.8	17
Street with trees	0.6	16
Street with higher number of trees	0.3	16

5.4 Results of the Simulations of the Urban Modules

The results of the simulations are listed at the end of this section. The values are rounded to the first digit after the decimal point.

The simulation results show that full cutoff and semi-cutoff luminaires emit the same amount of light towards the sky and the same luminance levels on the roads. Full cutoff luminaires primarily direct the light onto the street, and due to the lack of vertical surface illumination the urban space appears darker than streets with semi-cutoff luminaires (see Sect. 4.7). Semi-cutoff luminaires emit part of the light to the facades, making the space look brighter compared to full cutoff luminaires. The "darker" space generated by the full cutoff luminaires suggests lower light emissions, making the luminaires appear more "light pollution friendly". The simulation results refute this.[1]

The simulations implemented confirm that the proportion of light emitted by street luminaires decreases the higher the buildings become. In the EU models (building height 25 m), for example, they account for 18 and 22% of total emissions in the direction of the sky, and in the USA models (building height 180 m) for only three and one per cent. The same is true for shop windows, with EU models at 16 and 20% and US models at 1%.

Consequently, the facades with their reflection and absorption properties and the backlit windows play a major role in the overall emission. In this simulation constellation they also have the highest values at 58, 66, 96 and 98%.

The measured values also show that it is important to adapt the lighting to align with the cross-sections of the roads. It can clearly be seen that streets with narrower cross-sections have higher values than streets with wider cross-sections (see Sect. 4.7).

The percentages of total light emissions resulting from the simulations are used to evaluate the average luminance values in each category in Chap. 4. The percentage of light emission is used as a correction factor. Multiplying the average values determined in Chap. 4 by the respective correction factors results in a new value, which could be referred

[1]It should also be mentioned that the enhanced glare control provided by the full cutoff luminaires has a positive impact for insects (see Sect. 3.3): the light source is not so apparent to insects and is thus less attractive.

to as “weighted urban luminance” (guL = “gewichtete urbane Leuchtdichte”). This procedure allows all the individual values in the respective categories to be added and weighted accordingly. The metropolises would be allocated a total value to be referred to as “weighted total urban luminance” (guG = “gewichtete urbane Gesamtleuchtdichte”). This makes it possible to classify the individual metropolises. After further city analyses, it will be possible to create a catalogue containing different guideline values.

In the course of this, the respective simulation models/urban modules have been categorised according to the metropolises under study in Chap. 4. This results in the following overview:

- New York City
  Main roads = USA Main Road
  Side roads = USA Side Road
- Boston
  Main roads = EU Main Road
  Side roads = EU Side Road
- Toronto
  Main roads = USA Main Road
  Side roads = USA Side Road
- Warsaw
  Main roads = EU Main Road
  Side roads = EU Side Road
- Hanover
  Main roads = EU Main Road
  Side roads = EU Side Road

Below is an example calculation based on New York City. All results of the metropolises are listed in Table 5.3.

Example calculation:

The average luminance values (see Sect. 4.2) determined for New York City are:

Main roads

Facade:	1.9 cd/m^2
Shop window/entrance:	5.3 cd/m^2
Street:	1.0 cd/m^2
Sidewalk:	1.1 cd/m^2

Side roads

Facade:	0.8 cd/m^2
Street:	0.7 cd/m^2
Sidewalk:	1.1 cd/m^2

Table 5.3 Weighted luminance results

	Main roads guL				Side roads guL			
Metropolis	Facade	Shop/ent.	Street	Sidewalk	facade	Street	Sidewalk	guG
NYC	1.82	0.05	0.02	0.02	0.78	0.01	0.01	2.71
Boston	1.16	0.7	0.37	0.35	0.86	0.23	0.27	3.94
Toronto	1.44	0.03	0.05	0.05	1.57	0.02	0.01	3.17
Warsaw	3.95	3.36	1.12	1.45	2.90	0.55	0.59	13.92
Hanover	0.64	0.28	0.18	0.07	0.2	0.05	0.05	1.47

These values are multiplied by the respective correction factors (percentage of total light emission, Table 5.4, last column on the right). The categories street and pavement/sidewalk are multiplied by the same correction factor as the street lamp. This results from the fact that these surfaces are illuminated using the same light source. The result is:

Main roads

Facade:	1.9 cd/m^2 × 0.96 = 1.82 guL
Shop window/entrance:	5.3 cd/m^2 × 0.01 = 0.05 guL
Street:	1.0 cd/m^2 × 0.02 = 0.02 guL
Sidewalk:	1.1 cd/m^2 × 0.02 = 0.02 guL

Side roads

Facade:	0.8 cd/m^2 × 0.98 = 0.78 guL
Street:	0.7 cd/m^2 × 0.01 = 0.01 guL
Sidewalk:	1.1 cd/m^2 × 0.01 = 0.01 guL

The total value for New York City is thus 2.71 guG.

As a result of the findings, Warsaw produces by far the most light pollution, followed by Boston, Toronto, New York City and Hanover. At this point, however, it should be noted that the side streets in New York City were not located in the lighting hotspots as was the case in the other metropolises. As described in Sect. 4.2.2, no differentiations between main and side roads could be made in the New York City lighting hotspot. In order to be able to provide a representative value for New York City's lighting hotspot area, the main streets are rated as double that of the side streets. This results in a new value of 3.82 guG for the metropolis. This value is more realistic and the valid list is now:

1. Warsaw
2. Boston
3. New York City
4. Toronto
5. Hanover

Table 5.4 Results of the simulations of the urban modules

Simulation model	Scene	Average illuminance (lx) into the sky	Percentage of total light emission (%)
USA main road	Street luminaires (full cutoff), shop window lighting and facade windows on	7.7	100
	Only street luminaires on (full cutoff)	0.2	3
	Only shop windows on	0.1	1
	Only facade windows on	7.4	96
	Street luminaires (semi-cutoff), shop window lighting and facade windows on	7.7	100
	Only street luminaires on (semi-cutoff)	0.2	3
	Only shop windows on	0.1	1
	Only facade windows on	7.4	96
USA side road	Street luminaires (full cutoff), shop window lighting and facade windows on	8.1	100
	Only street luminaires on (full cutoff)	0.1	1
	Only shop windows on	0.1	1
	Only facade windows on	7.9	98
	Street luminaires (semi-cutoff), shop window lighting and facade windows on	8.0	100
	Only street luminaires on (semi-cutoff)	0.1	1
	Only shop windows on	0.1	1
	Only facade windows on	1.9	98
EU main road	Street luminaires (full cutoff), shop window lighting and facade windows on	7.6	100
	Only street luminaires on (full cutoff)	1.7	22
	Only shop windows on	1.5	20
	Only facade windows on	4.5	58
	Street luminaires (semi-cutoff), shop window lighting and facade windows on	7.6	100
	Only street lamps on (semi-cutoff)	1.7	22
	Only shop windows on	1.5	20
	Only facade windows on	4.5	58

(continued)

Table 5.4 (continued)

Simulation model	Scene	Average illuminance (lx) into the sky	Percentage of total light emission (%)
EU side road	Street luminaires (full cutoff), shop window lighting and facade windows on	9.2	100
	Only street luminaires on (full cutoff)	1.7	18
	Only shop windows on	1.5	16
	Only facade windows on	6.1	66
	Street luminaires (semi-cutoff), shop window lighting and facade windows on	9.2	100
	Only street luminaires on (semi-cutoff)	1.7	18
	Only shop windows on	1.5	16
	Only facade windows on	6.1	66

It should be noted that the simulation models were created based on the "shop windows/ entrances" category (window rows at 460 lm). This was done with foresight in preparation for Chap. 7. The "shop windows/entrances" category does not exist in the side roads in the comparative analysis (Chap. 4) and has therefore not been included in the calculation of the guL and guG.

It should also be noted that the weighted values do not necessarily represent the light pollution problem per se. The influencing factors in every city are of different origins. For example, all major cities have high facade values, but in Warsaw this is the result of clutter, and in New York City the result of backlit windows. The weighted results must therefore be explicitly compared based on the experience gained through the local analysis. Only then is it possible to identify the main sources of light pollution and find efficient solutions.

The simulation process can be refined by using more detailed models for each city.

In Google Earth most cities of the world are clearly depicted in detail (Fig. 5.10). These models were created based on measurements using satellite data and with the aid of satellite images. The models are depicted in such detail that even the vegetation or small bus stations are visible. In addition, all objects have textures, so you can include precise material properties in the models in the light calculations. All that has to be created for the individual buildings in the area under study are user profiles. These include how long and how much an office space is illuminated at night and how much light this emits into the urban space. Furthermore, it would be possible to integrate a Light Pollution plug-in into the Google Earth Engine. Geotagging could be used to integrate the images taken on site directly into the Google Earth model.

Unfortunately, Google does not provide the models right now. It is possible to export geometries from Google Earth through the third party program "3D Ripper DX". However,

Fig. 5.10 Example 3D of model Brunswick. (Source: Google Earth)

this does not lead to a satisfactory result (cf. Deep Shadows 2015). The post-processing of the models in order to be able to carry out light calculations is too time-consuming.

If this changes in the future, then this would be an efficient and fast way of working. This would include both on-site observations and actual emissions. Especially the observations on site need to be taken into account. To date this has been neglected in terms of light pollution.

Only by visiting towns is it possible to identify the many factors that cause light pollution in individual cities.

At present, the usual methods consist of comparing sky brightness with each other, as was the case with Cinzano et al. (2001, see Sect. 3.1). Further methods for monitoring light pollution can be found on the website http://lightpollution.pk.edu.pl/Eng/metody.php. These methods and approaches are all based on similar principles.

To date the methods focus only on the light emissions and do not investigate the causes (e.g. luminaires, surfaces). This has resulted in "blanket" solutions, such as the proposal to use full cutoff luminaires. The overall results of the analyses (see Chaps. 4 and 5) show that a "blanket solution" is not the answer. Every metropolis has different light pollution issues to deal with.

Bibliography

Cinzano, P./F. Falchi/C. D. Elvidge, "The first World Atlas of the artificial night sky brightness", in: *Monthly Notices of the Royal Astronomical Society* Vol. 328, H. 3 (2001), pp. 689–707.

City of Boston, LED Street Lighting, 31.07.2015, (accessed 10.01.2017). http://www.cityofboston.gov/publicworks/lighting/led.asp

Deep Shadows, 3D Ripper DX, 20.08.2015, (accessed 10. 01.2017). http://www.deep-shadows.com/hax/3DRipperDX.htm

Emporis, Boston Buildings, 31.07.2015a, (accessed 10. 01.2017). http://www.emporis.com/city/101045/boston-ma-usa

Emporis, Hanover, 06.08.2015b, (accessed: 10. 01.2017). http://www.emporis.com/city/100570/hanover-germany

Emporis, Skyscrapers in New York City, 16.08.2015c, (accessed 10. 01.2017). http://www.emporis.com/city/101028/new-york-city-ny-usa/type/skyscrapers

Emporis, Skyscrapers in Toronto, 16.08.2015d, (accessed 10. 01.2017). http://www.emporis.com/city/100993/toronto-canada/type/skyscrapers

Emporis, Toronto Buildings, 02.08.2015e, (accessed 10. 01.2017). http://www.emporis.com/city/100993/toronto-canada

Emporis, Warsaw Buildings, 04.08.2015f, (accessed 10. 01.2017). http://www.emporis.com/city/100803/warsaw-poland

Selene NY, Legislation, 29.07.2015, (accessed 10.01.2017). http://selene-ny.org/legislation/

The International Dark-Sky Association, Find Dark Sky Friendly Lighting, 16.08.2015, (accessed 10. 01.2017). http://darksky.org/fsa/fsa-products/

Vereinigung der Sternenfreunde e. V. Fachgruppe Dark Sky, List of fully shielded luminaires, 16.08.2015, (accessed: 10. 01.2017). http://www.lichtverschmutzung.de/seiten/vollabgeschirmt.php

6 Solutions for Reducing Urban Light Pollution

This chapter summarises the findings and conclusions of the previous chapters and uses them to develop solutions. The chapter is divided into three parts: optimisation of the collected findings on light pollution, the development of a roadmap/plan of action, and design guidelines plus the fictitious implementation of the solution concept taking Warsaw as an example.

6.1 Optimisation Based on the Knowledge Gained on Light Pollution

In the previous chapters, insights have been gathered into the different effects of light on urban space and the ecosystem. The conclusions drawn from this serve as a basis for optimising individual components so that less light pollution occurs.

In this context, a further column has been added to the summary tables in the previous chapters: "Optimisation with regard to light pollution". The tables are listed below. Further comments and explanations can be found in Sect. 6.2.

Optimisation focuses on ensuring that improvements are not subjected to interference. There are also "compromise solutions". An example of this is the effect of different light spectra on birds and insects. Insects are more attracted to the bluish light than to the yellowish light (see Sect. 3.3). Optimisation in this case, for example, would mean opting for a light spectrum without a blue component. However, the blue portion can have advantages for birds and their orientation in the urban space (see Sect. 3.2). The solution could therefore have negative consequences for the bird world. The "compromise solution" in this case is a light spectrum with a low blue component (warm white LEDs). The spectrum is less attractive for insects, but the orientation of the birds is still guaranteed.

E. E. Goronczy, *Light Pollution in Metropolises*,
https://doi.org/10.1007/978-3-658-29723-7_6

Below you will find tables summarizing what light pollution optimisation can be achieved for humans (Table 6.1), birds (Table 6.2), insects (Table 6.3) followed by a comparative analysis (Table 6.4).

Table 6.1 Light pollution optimisation for human beings

Cause	Impact	Light pollution optimisation
Glare produced by the light source	• Disncomforting, disturbing effect • Impairs visual performance	• Light source needs to be shielded • Attention must be paid to luminaire positioning and mounting height
LED spectrum (high blue component)	• Unnaturally bright effect • Extends adaptation time • Suppresses melatonin production	• Avoid using cool white LEDs, opt for warm white LEDs
Brightening of night sky (sky glow)	• Suppresses the secretion of different hormones in the metabolic system, including melatonin, which can promote: – Diabetes – Overweight – Cancer – Myocardial infarction	• Optimisation of light emission by providing solutions that are appropriate for the respective towns

Table 6.2 Light pollution optimisation for birds

Cause	Impact	Light pollution optimisation
Light source	• Birds are attracted by luminaires, fly into the housings and are killed immediately • Rays of light attract birds, leading to mass collisions • Light interferes with birds' sense of orientation	• Skybeamers should be forbidden • Lighting concepts for interiors of buildings; avoid exterior lighting on buildings • Avoid excessive facade illumination • Facade lighting should be affected from the top downwards (downlights)
Light spectrum (yellowish)	• Can interfere with birds' sense of orientation	• Opt for warm white LEDs
Brightening of night sky (sky glow)	• Orientation by means of the stars is inhibited • Disruption of the circadian rhythm, which affects/undermines: – Hunting behaviour – Breeding seasons – Courtship behaviour – Physical condition • Suppresses the secretion of melatonin and thus disrupts the time-associated orientation of migratory birds	• Optimisation of light emission by providing solutions that are appropriate for the respective towns

Table 6.3 Light pollution optimisation for insects

Cause	Impact	Light pollution optimisation
Light source	• Insects are attracted by the light source, and as a result they: – Use more energy flying around – Have less time to search for food and for reproductive activities – Are lured away from their natural habitat – Can be killed or burnt by the light source – Become easy prey for their predators	• Uniform illumination. • Light sources need to be shielded well • Sealed luminaire housing.
LED spectrum (high blue component)	• Highly attractive to insects. • Suspected to attract new insect species that are sensitive to short-wavelength radiation	• Opt for warm white LEDs
Brightening of night sky (sky glow)	• Interrupts the insects' circadian rhythm • Negative impact on hatching times • Orientation by means of the stars is inhibited	• Optimisation of light emission by providing solutions that are appropriate for the respective towns

Table 6.4 Light pollution optimisation—a comparative analysis

Urban modules/ properties	Impact	Light pollution optimisation
Facade surface (material properties) Building density Building height Light spectrum of luminaire Prevalent average level of brightness	• Different urban spaces are perceived as being brighter or darker although the measured values are the same	• Depending on the desired spatial effect materials can be applied that reflect the light accordingly, dull surfaces make for soft diffuse light • Adjust the luminous flux in accordance with the building density: the higher the building density, the lower the luminous flux • Use vertical surfaces to illuminate the urban space with little luminous flux • Aim to achieve an overall low level of brightness
Cross-section of road	• The same lighting scheme will look brighter if the cross-section of the road is narrow, and darker if it is wide	• Adjust the luminous flux of the luminaires to align with the cross-section of the road; narrow cross-section roads require less luminous flux

(continued)

Table 6.4 (continued)

Urban modules/ properties	Impact	Light pollution optimisation
Facades of skyscrapers/ high-rise buildings	• Given the illuminated windows, facade illumination and reflective facade surfaces, extensive facades have great light pollution potential	• Adjustment of switching times • Lighting concepts for interiors of buildings; avoid exterior lighting on buildings • Avoid excessive facade lighting • Facade lighting should be affected from the top downwards (downlights) • Smart streetlights that adapt to the level of brightness required
Illuminated vertical surfaces	• Disproportionate light emission/ light pollution	• Avoid excessive facade lighting • Facade lighting should be affected from the top Downwards (downlights)
Mounting height Position of luminaire Luminaire spacing Light distribution curve	• Determines level of uniformity of the lighting • Determines degree of clutter and light trespass	• High mounting heights for uniform lighting of the space and glare control • Position luminaires, taking advantage of the features in the urban space, e.g. combination of luminaires and trees • Opt to apply agreed light distribution curves, pole spacing and number of luminaires to avoid clutter and light trespass
Luminaire-tree combination	• Good lighting with little light trespass into the sky	• Taking advantage of the features in the urban space
Combination of shop windows and luminaires	• Danger of creating a clutter situation	• Smart streetlights that adapt to the level of brightness required
High level of brightness on main roads	• No lighting required in side roads in certain areas	• Only install street lighting where it is truly needed
(High volume of traffic)	• (The urban realm is very bright)	• Smart streetlights that adapt to the level of brightness required

6.2 Development of a Roadmap and Design Guidelines

The roadmap consists of four stages or steps: Analysis, Measures to be taken Part 1, Measures to be taken Part 2, and Measures to be taken Part 3 (Fig. 6.1).

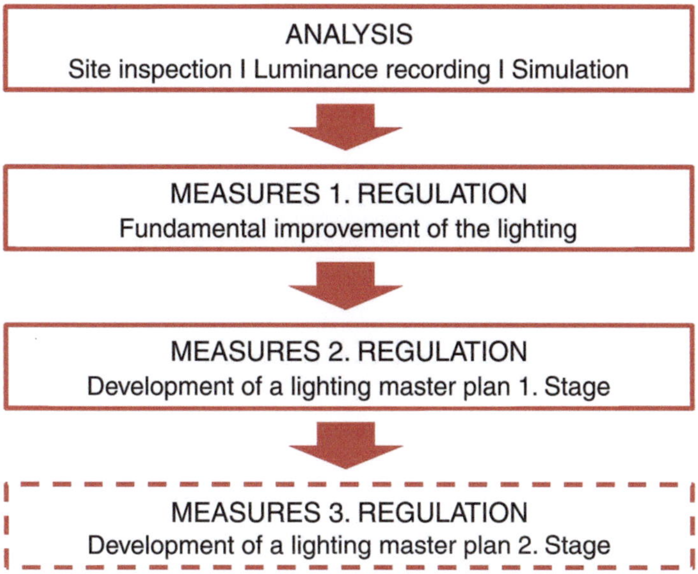

Fig. 6.1 Roadmap

Before light pollution-reducing measures can be considered, an analysis of the location is essential. This should be carried out in accordance with the procedure already described. This ensures a careful analysis of the main causes of light pollution.

The measures to be taken are arranged according to the costs they can incur. They all comprise a master plan, each of which addresses different issues.

The Measures to be taken Part 1 incur the lowest costs and expenditure. The Measures to be taken Part 3 incur the highest costs and are generally only feasible in new parts of a town/city which are still in the planning stage. The list of measures to be taken can be regarded as an additive process. That is to say that the Measures to be taken Part 3 include the measures undertaken in the previous two stages, following the analysis. The different stages serve primarily as an orientation aid; the measures defined in the different stages can also be combined. The measures to be taken that are listed are based on the experience gained in the previous chapters and summarised in the tables in Sect. 6.1.

The steps that make up the roadmap are explained in more detail below.

Analysis

The analysis consists of the elements and procedures that have been explained and carried out in this work in Chaps. 4 and 5. To sum up, the first step is to carry out a site inspection to take note of the lighting equipment in the respective city or district to be improved. The results should include: an inventory of the luminaires (light source, mounting height, glare, etc.), a description of the impressions gained on site (urban space, lighting scenario, etc.), plus luminance images and their evaluation.

The site inspection is extremely important. The results can already be used to identify fundamental issues such as excessive brightness due to clutter as seen in Warsaw (see Sect. 4.5.2).

The evaluations of the luminance images are used to create 3D lighting simulations. The results illustrate the impact of the individual categories in per cent. Light emissions in the area become visible and are compared with the experience made on site. The main cause of light pollution becomes clear.

In the simulation, an average model can be used for the area as was the case in Chap. 5. If detailed models are available for the area under study, these can be used to obtain more accurate values.

The analysis is the most important component. A sound end result can only be achieved given a solid basis.

On the basis of the analysis, individual problems and issues can be defined and the measures to be taken planned accordingly.

Measures to Be Taken Part 1

The Measures to be taken Part 1 are basic measures to minimise light pollution. They comprise low expenditure and investment costs. The solutions addressed require no structural measures to be taken. The space is optimised only in terms of lighting technology. This does not necessarily improve the quality of stay in the urban spaces, but it does not worsen it either. The master plan is used only for analytical purposes. The measures to be taken are considered from an economic point of view. The measures include:

- *Replacement of conventional lamps with warm white LEDs*

  The warm white spectrum of the LED has a low blue component and no UV component. As explained in Sect. 6.1, this is a "compromise solution" from which all parties can benefit. From a biological point of view, the optimum light source for road lighting for insects and humans is a sodium vapour lamp with a PMMA diffuser (see Sects. 3.1 and 3.3). For birds, however, lamps with a high blue and green component (see Sect. 3.2) or even monochrome blue or green light are specified (cf. Poot et al. 2008).

  LEDs are also highly efficient light sources, which brings further ecological advantages.
- *Reduce overall level of brightness*

  The overall brightness of a city determines the sky glow in the area. Comparisons between cities have shown that it is still possible to ensure orientation and a sense of security in the urban space even with low levels of brightness. The illuminance in the urban space needs to be measured, and the luminaires dimmed and lamps replaced or even removed accordingly.

In addition to reducing light emissions, reducing the level of brightness in the urban space also offers great energy-saving potential.

- *Optimisation of switching times*

  The switch-on and switch-off times should be adapted to the seasons or sunrise and sunset. In many cities, it is already common to use summer and winter on/off times. In the future, this must be done with even more precision. It is possible today to calculate the sunrises and sunsets for every single day over decades in advance (Sunrise and Sunset 2015).
- *Adjustment of luminaires*

  The adjustment of the luminaires must be checked in the areas under study and corrected if necessary. Luminaires that are not directed or adjusted accurately can give rise to glare, clutter and light trespass.
- *Prohibition of skybeamers*

  Skybeamers do not enhance the urban space, nor do they deliver the right light for urban spaces. For this reason, the light emitted by skybeamers can be considered as 100% light pollution. These luminaires have a fatal effect on birds (due to the light beam, cf. Sect. 3.2) and on insects (due to the light beam and the luminaire design, cf. Sect. 3.3).
- *Prohibition of advertising spaces that are illuminated*

  The analysis in Chap. 4 shows that especially advertising surfaces that are illuminated have high light emissions as a result of the reflected light. Advertising spaces should not be additionally illuminated; the urban lighting is sufficient. Advertising space in the urban space after dark already attracts a lot of attention due to size. Additional lighting is therefore not necessary.

Measures to Be Taken Part 2

The Measures to be taken Part 2 and Part 3 are focused on the creation of a constructive and creative lighting master plan. This requires the involvement of a qualified lighting designer. The solutions are combined with structural measures that give the urban space a quality of stay at night. Design aspects play an important role in this.

The proposed design concepts developed as part of the Measures to be taken Part 2 and Part 3 should not be subject to the basic conventional understanding of day and night. A cultural rethink must take place.

The urban spaces do not have to be illuminated at night to the extent that they have the same appearance at night as during the day. The urban space should have a different face, a different character, or a different effect at night from during the day. This principle can be illustrated symbolically through glowworms. Glowworms are rarely noticed during the daytime and are regarded by many as being unsightly, or ugly even. After dark, their appearance changes. They start to glow and have a nocturnal aesthetic that fascinates many people. A good lighting master plan, or a good lighting design, should achieve the same effect and reduce light pollution into the bargain.

When designing a lighting master plan to minimise light pollution, it is important to ensure a certain level of uniformity (cf. tables in Sect. 6.1) and a basic low level of brightness (cf. tables in Sect. 6.1) in the urban spaces. These comprise different elements which can be given different priorities, which in turn need to be addressed by the lighting design. Furthermore, in city streets it is essential to illuminate the facades very softly (at least the lower floors). The urban space should still be visible as such at night—not only for the purpose of orientation, but also to promote a sense of security. Full cutoff luminaires, which are designed to make the room appear dark, must be used with caution.

In terms of design, the Measures to be taken Part 2 concentrate primarily on public spaces and the luminaires installed there. The Measures to be taken Part 3 takes this a step further, and addresses also "semi-public" elements such as office buildings or cultural buildings.

A good example of the illumination of public urban space is the High Line in New York City (Figs. 6.2, 6.3 and 6.4). The basic uniform spread of brightness is created by the urban environment (skyscrapers, lighting in surrounding streets). Individual elements are accentuated to enhance the atmosphere, but also for orientation. Many aspects that prevent light pollution are to be found here. For example, the luminaires installed are almost exclusively all directed downwards, so that only very little light can enter the atmosphere (due to surface absorption). The luminaires are positioned below shoulder height when people are seated. Glare from the light source is therefore excluded.

On the whole, the High Line features a very low level of brightness, which enables it to act as a backdrop in the urban context. It overlooks the brightly lit city, which thus functions as a stage.

Fig. 6.2 The High Line, ground plan. (Source: Hervé Descottes, Architectural Lighting Designing with Light and Space, p. 85)

Fig. 6.3 The High Line, sectional drawing. (Source: Eric Laignel, Architectural Lighting Designing with Light and Space, p. 86)

Fig. 6.4 The High Line in the evening. (Source: Iwan Baan, Architectural Lighting Designing with Light and Space, p. 129)

Fig. 6.5 Atjehstraat, Rotterdam. (Source: Rudolf Teunissen, Marinus van Der Voorden Hans Wilschut, Lieselot IJzendoorn, Thoams Linders Max, accessed: 11.02.2018. https://www.brinklight.sg/blog/2012/07/broken-light-in-rotterdam/)

The result is an urban space that produces little light pollution yet provides a highly attractive place, with its own night-time identity, that people enjoy spending time in.

Another good example is the Atjehstraat in Rotterdam (Fig. 6.5). The sidewalk is illuminated via a pattern of light and shadows, giving the space an unmistakable night-time feel. In addition, the monolithic structure of the buildings is dissolved at night by the strip-like illumination of the facades after dark. The light stripes also influence people's perception of the street space. The urban space as a whole features a low level of brightness due to the shadow elements.

"The Gherkin" (London) and the Mori Tower (Tokyo) skyscrapers are prime examples of the semi-public space.

The Gherkin (Fig. 6.6) proves that urban architectural icons do not have to be excessively illuminated in order to attract attention. The building is illuminated from the inside, which results in comparatively little light emission. In the daytime, the building has a smooth surface. The lighting design after dark renders the outer skin translucent. The building thus features stronger ornamentation and spatial depth at night, which makes the building stand out in the urban space. Details that are not visible during the day become visible at night. Individual floors and the depth of the diamond-shaped skeleton become perceptible through the artificial light.

The lighting design of the Mori Tower in Tokyo is particularly ecologically significant (Fig. 6.7). There are four light scenes that adapt to ecological conditions, such as bird

Fig. 6.6 "The Gherkin". (Source: Diego Delso, accessed: 11.02.2018. https://upload.wikimedia.org/wikipedia/commons/a/af/Gherkin%2C_Londres%2C_Inglaterra%2C_2014-08-07%2C_DD_053.JPG)

Fig. 6.7 Different lighting scenes on the Mori Tower. (Source: Lighting Planners Associates (LPA), Professional Lighting Design Magazine, No. 91, p. 39)

migration. Here, too, the lighting design creates a night-time identity for the building without generating high light emissions.

Mondeal Square in Ahmedabad (Fig. 6.8) is an example of how semi-public and public spaces meet via good lighting design.

The subtle and required basic brightness is achieved using luminaires with reflective surfaces. The nocturnal identity is created by the buildings glowing from the inside coupled with the illuminated elements in the foreground. The principles are similar to those of the High Line.

In comparison, poor lighting solutions are shown in Figs. 6.9, 6.10 and 6.11. These images represent common scenarios of our time.

In these spaces there are no elements accentuated by light, and vertical surfaces are illuminated "randomly". The urban spaces are "robbed" of their architectural and spatial quality at night.

Purposefully designed lighting makes it possible to create urban identities while reducing light pollution. The skylines of the various metropolises have unique design potential.

Many large metropolises have skylines that act as landmarks for them. The skyscrapers that form the skyline are often floodlit to a high degree. Thus they have almost the same appearance at night as during the day. The illumination of the facades results in high light emissions, which can be seen in the satellite images in Chap. 4.

A lighting master plan for an entire metropolis offers numerous advantages from which the skylines in particular can benefit.

The lighting master plan reduces the overall level of brightness in the city and consequently also that of the skyline. This is not a disadvantage, because a "dark" skyline opens up new opportunities for designing. Viewed in the daytime, a skyline consists of different buildings of different shapes. The buildings are generally not subject to a uniform design principle. Thus the daytime appearance of a skyline is "disorderly".

Lighting design can counteract this. Artificial lighting can be applied to create a skyscraper scenario, allowing the shapes of the high-rise buildings—which are randomly assembled during the day—to come together at night to form a new composition (Fig. 6.12).

The radical design in Fig. 6.12 shows a potential redesign of the skyline in Shanghai by night. The skyline in Shanghai is especially chaotic during the day due to the unconventional shapes of the many skyscrapers. This confusing backdrop is currently suffering even more, because the buildings are on the one hand extremely brightly lit, and on the other because most skyscrapers have their own dynamic lighting design, which is not coordinated with surrounding structures (see Fig. 2.1). The skyscrapers "are fighting" with one another using light.

In the fictitious design, the overall level of brightness is reduced to a more humane level. The dome of light over the city disappears. The starry sky above the metropolis becomes visible again. The individual lighting concepts that generate light pollution are replaced by a holistic skyline concept. Only the vertical edges are illuminated with warm white light.

Fig. 6.8 Mondeal Square. (Source: Atelier dada, Neha Mevada, Pratik Chandresha, Professional Lighting Design Magazine, No. 98, title page)

Fig. 6.9 Petronas Towers. (Source: Derrick Caluag,, accessed: 11.02.2018. https://www.flickr.com/photos/eunick/5313131373/)

Fig. 6.10 Warsaw. (Source: author)

Fig. 6.11 Leicester Victoria Park. (Source: Ashley Dace, accessed: 11.02.2018. http://www.geograph.org.uk/photo/2731144)

Fig. 6.12 Fictitious lighting design for the Shanghai skyline

What is between the vertical lines moves into the background, revealing a grass-like landscape of lines of light.

A visibly coherent skyline at night has a more striking impact than numerous individually lit buildings. Skylines with their own night identity can thus be designed for cities that differ significantly from other skylines. New iconic representations of cities can be created, while at the same time protecting the environment from excessive light spill.

In addition to the design principles and guidelines, the Measures to be taken Part 2 also comprise the following issues:

- *Use luminaires with appropriate light distribution curves and luminous flux*

  Luminaires and light distribution curves should be selected/determined to ensure the facades are only illuminated to a minimum. Spaces with illuminated vertical surfaces are perceived as being brighter (see Chap. 4). This means that well lit spaces can be created with low luminous flux levels. In addition, softly illuminated facades make the space appear more uniform and visually comprehensible, which facilitates orientation.

  The light distribution curves and luminous flux levels should be determined according to the requirements of the given urban space and its cross-sections. On-site testing is recommended.
- *Install high-quality luminaires*

  Glare can be avoided using well designed luminaire. The luminaire head should be sealed to prevent ingress of dirt and small creatures such as insects.
- *Smart lighting*

  "Smart" luminaires are equipped with sensors that can detect brightness and movement. These luminaires are therefore extremely efficient. The luminaires adjust the luminous flux according to the brightness of the surroundings. There are also lights that dim down when there are no people in the street. Using these luminaires, any overexposure in the form of clutter is prevented in urban spaces.
- *Optimum lighting layout (positioning of luminaires)*

In the area to be designed, the existing positions of the luminaires must be checked. The audit should also address the urban potential to benefit from purposefully designed lighting effects (for example, the combination of trees and luminaires). This detailed analysis of the area enables the number of luminaires, positions of the luminaires, pole spacing, mounting height and luminous flux to be optimised for the respective site. Clutter and light trespass are thus avoided in the planning process.

Measures to Be Taken Part 3

As explained above, these measures are more far-reaching and can also affect the semi-public sector. The measures again also reduce light pollution. The identity of the locations/parts of the city is thereby increased.

- *Lighting design of the buildings*

  Light emission is further reduced if the surrounding buildings also feature "anti-light pollution" lighting design. It is recommendable to plan the lighting so that the outer appearance of the building is generated by light from the inside (see examples in Measures to be taken Part 2, Figs. 6.6, 6.7 and 6.8). In the case of facade lighting, this should be realised using luminaires directed towards the ground.

  For offices, it is possible to equip the luminaires with an intelligent control system. Accordingly, the light is only switched on in those parts of the building where people are present. Consequently, the office structure needs to be rethought. Workstations that are frequently occupied after dark must be oriented towards the core of the building. These measures minimise light trespass into the urban space.

  Office space where it is already clear in advance that lighting will mainly be required after dark should be planned in on the lower floors. This enables more light to be absorbed by the urban surfaces, thus emitting less light into the sky.

  The working times in some buildings can also be optimised. Cleaning staff, who often work after normal office hours and have to illuminate whole floors, can postpone the work to the daytime (in the employees' lunch break or early in the morning).
- *Luminaires designed specifically for the location*

  Depending on the site to be worked on, luminaires can be developed which are adapted to the location both in terms of lighting technology and design. This increases the quality of stay as well as enhancing the identity of the space. An example of this is the beach promenade in Niendorf (Fig. 6.13).
- *Lighting design for new buildings that are in the planning stage*

  In projects where the lighting designer is involved in the planning right from the start, the best results can be expected. All the factors mentioned above are taken into account during the planning phase. Furthermore, by determining the surface materials in the space, it is possible to create the desired spatial effect and enhance the quality of stay even more profoundly.

Fig. 6.13 Niendorf promenade. Source: Studio DL Germany, accessed: 10.01.2017

To sum up, it can be said that the Measures to be taken Part 1 can already make a major contribution to combatting light pollution.

However, if a town or city seriously wants to tackle the issue of light pollution, a lighting designer must be involved.

The Measures to be taken Parts 2 and 3 can entail high investment costs. However, the benefits of a good lighting master plan must not be lost sight of:

- protection of the environment by minimising light pollution
- promotion of urban quality of stay and identity of a space
- energy savings and cost reduction

6.3 Fictitious Implementation of the Solution Concept Taking the Warsaw Metropolis as an Example

The procedure for minimising light pollution presented here is illustrated taking the City of Warsaw as an example. In this example, the overall light pollution of the city is to be reduced. The site as documented in photo session locations 11–18 NW (Fig. 6.14) will be examined more closely and further optimised with regard to light pollution.

Fig. 6.14 Area to be optimized

Area Analysis of the Entire City

The detailed analysis of the City of Warsaw can be found in Sect. 4.5. The problems recorded in Warsaw are: clutter, light trespass and the excessive illumination of vertical surfaces.

Clutter is caused by the sometimes very dense spacing of the light poles and the number of luminaires at one point. The side streets are ultra-bright because they follow the same lighting principles as the main streets, but have a narrower cross-section. In some streets high levels of brightness are caused by luminaires positioned close to bright shop windows.

Light trespass occurs mainly in facade lighting. Light trespass caused by street lights, cannot be evaluated on site. The urban space is extremely bright, making it difficult to identify the impact of the individual luminaires on the overall situation.

All in all, it can be said that all facades in Warsaw are very bright.

Vertical surfaces such as facades or advertising surfaces that are purposefully illuminated have extremely high luminance values.

The simulations in Chap. 5 show that—at a guG value of 14—Warsaw emits a considerable amount of light. From the simulations it can be clearly seen that the greatest amount of light emitted towards the sky is generated by facades and roads.

After comparing all impressions collected on site, it can be established that the main cause of light pollution is the street lighting.

Measures

Simple measures are to be applied to ensure Warsaw produces less light pollution as a whole. Measures to be taken Part 1 comes into effect in this respect. In order of priority, these are the following actions recommended for the City of Warsaw:

- *Reduce overall level of brightness*
  Dim luminaires, or remove them if necessary. If dimming is not technically possible, a lamp with the appropriate luminous flux must be used. This effectively reduces the degree of brightness on the facade.
- *Adjustment of luminaires*
  Wrongly directed luminaires must be readjusted. This also helps to make the facades darker.
- *Prohibition of illuminated advertising space*
  The level of brightness in the urban space is sufficient to render advertising spaces legible. Further illumination of these surfaces causes unnecessary light emissions and makes the urban space even brighter than intended.
- *Replacement of conventional lamps using warm white LEDs*
  The lamps must be replaced successively. Warsaw's main problem is the excessive level of brightness in the urban spaces. The goal of replacing the light sources is to create a healthier light spectrum for the environment and the people. Warsaw uses sodium vapour lamps in the main (see Sect. 4.5.2). With regard to human beings and the environment, this spectrum has less negative impact (see Chap. 3). Consequently, this is of secondary importance when it comes to the first measures to be taken. Another action that can be considered is retrofitting conventional luminaires with PMMA diffusers. These absorb the UV range in the sodium vapour lamp spectrum, thereby improving it (see Sect. 3.3).
- *Optimisation of switching times*
  The stay in Warsaw was too short to be able to make statements about the existing switching times. Nevertheless, appropriately planned switching times are recommended and reduce light emissions at dusk and dawn.
- *Prohibition of skybeamers*
  Only one skybeamer was found on site. Overall, this factor therefore plays a small role. However, these luminaires have a high hazard potential for birds and insects, which justifies a ban in principle.

The listed catalogue of measures reduces light pollution across the entire area, which also minimizes the sky glow.

The area to be examined more closely in Fig. 6.14 comprises side roads. Here, light emissions need to be further reduced. In order to achieve this, Measures to be taken Parts 2 and 3 are applied.

Analysis of the Area Under Study

The area has the same main problems as described above. As these are side roads, however, there is no illuminated advertising space. Similar to on the main roads, the side roads also have cafés and shop windows on the lower floors.

Due to the narrower street cross-section, the side roads are generally brighter than the main roads. In this area, mainly light-coloured stone facades are to be found, which

Fig. 6.15 Location for photo session 16 NW

Table 6.5 Evaluated luminance results 11–18 NW

	streets guL				
metropolis	facade	shop windows/entrances	road	pavement	guG
Warsaw 11–18 NW	3.89	1.61	0.59	0,66	6,75

generates an impression of increased brightness. In the area under study, there is no accentuation of orientation elements or lighting that underlines the architecture. Because functional lighting has no relation to urban structures or architecture, many forms and structures are perceived as being lit in a standard—albeit unfavourable—way. The spatial quality that the place possesses during the daytime is thoroughly undermined by the night-time lighting (Fig. 6.15).

In order to evaluate the rated emissions in the area, the corresponding average values of the roads are multiplied by the correction factor of the EU Minor Street simulation model (Table 6.5).

The results show the same trends that can be found throughout the city. Light emissions occur in the main via the facades, followed by the shop windows/entrances category. Compared with the other metropolises in the analysis in Chap. 4, the area possesses a high guG.

At this point it should be mentioned that the building block models in Chap. 5 can mapped into a de facto light emission rating. An example of this is shown in the street section at photo session location 16 NW. The road has a length of approximately 250 m. The building blocks are 200 m long. This means it is the equivalent of 1.25 times (9.2 lx + 2.3 lx) the emissions recorded for EU Side Road building blocks.

In this case, the illuminance emitted in the direction of the sky is 11.5 lux. Based on these values, it will be possible in future to produce guideline values that allow the different illuminance levels to be assessed.

In order to further minimise light pollution, on the basis of the analysis Measures to be taken Parts 2 and 3 are applied.

Measures

The values in the facade and shop windows/entrances categories are particularly high in the area under study. In order to solve the problem of light pollution, the lighting master plan must include not only street lighting, but also semi-public buildings.

As an example, actions taken are shown in the street section at photo session location 16 NW (Figs. 6.16 and 6.17).

In the design, the pole-mounted lights at the sides of the road have been removed. The luminaires have an unfavourable light distribution curve which projects light onto the facades. This has an unfavourable effect on the structure of the building (Fig. 6.15, facade on the left). These luminaires are replaced by luminaires suspended from cables positioned in the centre of the urban space, thus ensuring minimal and uniform lighting of the space. The fact that the space is clearly visible facilitates orientation and supports the sense of

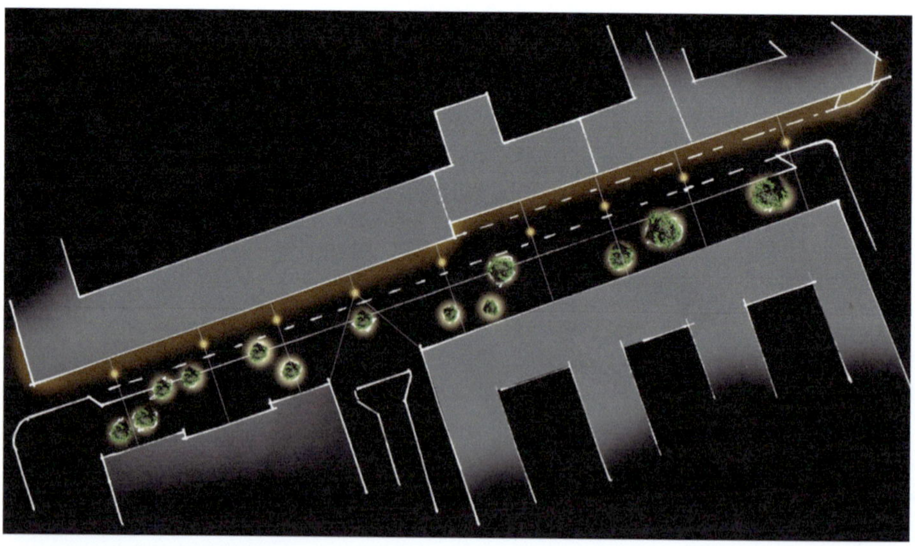

Fig. 6.16 Plan view of the design

Fig. 6.17 Sketch showing a perspective of the road

security. The illumination of the roadway provided by the luminaires is also standard-compliant.

The trees on the sidewalks are accentuated using uplights to provide additional orientation and give the room an identity. The tree trunks are pleasantly illuminated, an effect that also exists during winter months. In the summer, the advantages of the combination of trees and light come to the fore (see Sect. 5.3).

The quality of the architecture is highlighted by the new lighting design. This is achieved by ensuring the panes in the lower area and the window soffits in the upper area receive an even spread of light. As a result, the architecture is accentuated at night and this part of town is given an unmistakable "face".

In this way, architecture is not pushed into the background or all made to look equal, as is the case with the current standard illumination of urban spaces. The aesthetic quality of the architecture is accentuated with light and the urban space thus gains with regard to quality of stay at night. At the same time, less light is emitted. If the luminaires are correctly adjusted, hardly any light enters the atmosphere thanks to the trees, window soffits and panes. On the one hand, the light is not radiated directly towards the sky and on the other hand, through the positioning of the light sources a large part of the light is absorbed by the urban surfaces.

Given that they deliver a minimum level of brightness to the entire space, the luminaires suspended above the roadway emit the most light in the direction of the sky. The decisive factor here is the extent to which the level of brightness can be lowered in the overall area. Any improvement will be correspondingly high in comparison to the existing situation.

Another positive aspect of the design is the low glare potential due to the positioning of the luminaires.

In the case of the concept presented, it should also be noted that a lighting plan should be drawn up for the shop windows and cafés. These can destroy the overall impression of the urban space through excessive lighting.

Bibliography

Poot, H./J. E. Bruno/H. de Vries/M. Donners/M. R. Wernand/J. M. Marquenie, "Green Light for Nocturnally Migrating Birds", in: *Ecology and Society*, H. 2 (2008).

Sunrise and Sunset, Sunrise and Sunset, 21.08.2015, (accessed: 10. 01.2017). http://www.sunrise-and-sunset.com/de

7 Conclusion and Outlook

The subject of light pollution is gaining increasingly more presence. As explained in Sect. 3.1, global brightness after dark is increasing from year to year. The effects of this development are also becoming increasingly drastic. Different bodies such as industry, natural scientists or astronomers are trying to get the problem of light pollution under control.

So far, this has been restricted to "package solutions", which are not very efficient, and in most cases these solutions are to the detriment of the architecture or the quality of life in urban areas. Due to the increasing interest in the topic and the resulting willingness of municipalities and cities to act, more and more investments are being made in new lighting systems. On the one hand, these are intended to save energy, but on the other hand it is believed they should also produce less light pollution.

This trend should be seen as an opportunity to increase the quality of architecture and of life in urban spaces at night by means of a new lighting design approach.

In order to be able to implement this, however, a novel analysis method is required which can pinpoint and reveal where potential lighting problems lie. This is the only way to find an efficient solution to the problems. The concepts and ideas described in this book serve as a basis. The solutions developed should not only be aimed at curbing light pollution. The design aspect should not be overlooked when it comes to realising new lighting schemes.

Good lighting design can bring architecture to the fore again at night, enhance the quality of stay in urban spaces, and effectively combat light pollution. Unique night-time identities can be created for different quarters or even entire cities. In future, monotonously illuminated streets, which ignore architectural qualities and show no signs of enhancing the quality of stay, could really be a thing of the past.

One development emerging today in the lighting industry is the automation of luminaires (smart lighting). Luminaires adapt to the ambient brightness. Light pollution

E. E. Goronczy, *Light Pollution in Metropolises*,
https://doi.org/10.1007/978-3-658-29723-7_7

Fig. 7.1 Toronto 2003, during the blackout (left), Toronto with normal street lighting (right). Source: Todd Carlson, accessed: 10.01.2017 http://www.darksky.org/light-pollution/

caused by clutter will therefore occur only seldom in the foreseeable future. Using these luminaires, light pollution can be considerably minimised. That said, design aspects are not addressed, so the quality of the urban space is not necessarily enhanced by using "smart lighting".

It is highly desirable to continuously minimise light pollution. One huge advantage that can be drawn from this is that the urban realm can regain the natural conditions that have suffered or been lost through light pollution. In future, it will be possible to view the beauty of the stars, including the Milky Way, in metropolises at night instead of a veil of light.

Figure 7.1 shows a residential street in Toronto during a power failure in 2003 (left: no light pollution) compared to normal lighting conditions after dark (right: light pollution caused by inappropriately positioned luminaires and light trespass in the metropolis).

Annex

A.1 Measurement Data on Buildings in Metropolitan Areas as a Basis for Chap. 5

A.1.1 New York

Average building heights

Table A.1 New York City building heights

Building	Height (m)	Floors	Building type	Year
One World Trade Center	541	104	skyscraper	2014
432 Park Avenue	426	96	skyscraper	2015
Empire State Building	381	102	skyscraper	1931
Bank of America Tower	366	58	skyscraper	2009
Chrysler Building	319	77	skyscraper	1930
New York Times Tower	319	52	skyscraper	2007
10 Hudson Yards	310	52	skyscraper	2015
One57	306	75	skyscraper	2014
Four World Trade Center	297	72	skyscraper	2014
70 Pine Street	290	66	skyscraper	1932
The Dylan	138	34	skyscraper	2013
160 Madison Avenue	138	42	skyscraper	2015
Tower 45	137	40	skyscraper	1989
1065 Avenue of the Americas	137	38	skyscraper	1957
Metropolitan Life North Building	137	30	skyscraper	1933
Sovereign Apartments	137	48	skyscraper	1973
Tower 31	137	41	skyscraper	2006
47 East 34th Street	137	36	skyscraper	–

(continued)

E. E. Goronczy, *Light Pollution in Metropolises*,
https://doi.org/10.1007/978-3-658-29723-7

Table A.1 (continued)

Building	Height (m)	Floors	Building type	Year
Park Lane Hotel	137	46	skyscraper	1971
Dag Hammarskjold Tower	137	43	skyscraper	1984
667 Madison	101	24	skyscraper	1989
Archstone Brooklyn Heights	101	33	skyscraper	2000
Grand Millennium Condominums	101	32	skyscraper	1996
Chelsea Tower	101	33	skyscraper	2003
Octavia Condominiums	101	32	skyscraper	1984
2 Wall Street	101	22	skyscraper	1933
Avenue of the Americas Plaza	101	22	skyscraper	1990
Belmont Apartments	101	33	skyscraper	1981
150 East 57th Street	101	32	skyscraper	2000
292 Madison Avenue	100	26	skyscraper	1923
Mean value	**198**			

A.1.2 Toronto

Average building heights

Table A.2 Toronto building heights

Building	Height (m)	Floors	Building type	Year
First Canadian Place	298	72	skyscraper	1976
Trump International	277	57	skyscraper	2012
Scotia Plaza	275	68	skyscraper	1988
Aura	272	78	skyscraper	2014
TD Canada Trust Tower	261	53	skyscraper	1990
One Bloor	257	76	skyscraper	2015
Commerce Court West	239	57	skyscraper	1972
ÏCE II	234	67	skyscraper	2015
TD Bank Tower	223	56	skyscraper	1967
Bay Adelaide West	218	51	skyscraper	2009
Ocean Club, South Tower	131	39	skyscraper	2015
Parade, West Tower	131	39	skyscraper	2011
Nautilus	130	40	skyscraper	2012
Beyond the Sea, South Tower	130	37	skyscraper	2011
Empire Condominium	130	28	skyscraper	2005
Dundee Place	129	31	skyscraper	1991
Spectra	129	39	skyscraper	2015

(continued)

Table A.2 (continued)

Building	Height (m)	Floors	Building type	Year
Picasso	129	39	skyscraper	2015
Leaside Towers II	129	44	skyscraper	1970
Leaside Towers I	129	44	skyscraper	1970
WaterPark Place I	101	24	skyscraper	1986
James Cooper Mansion	100	32	skyscraper	2011
Carrington Tower Condominium	100	34	skyscraper	1977
The Residences at the Consilium 2	100	33	skyscraper	1991
The New Residences of Yorkville Plaza	100	31	skyscraper	1971
West Village 2	100	30	skyscraper	2014
City Hall East Tower	100	27	skyscraper	1965
Toronto Star Building	100	25	skyscraper	1971
Yonge Eglinton Centre II	100	22	skyscraper	1974
Nestlé Building	100	21	skyscraper	1994
Mean value	**162**			

A.1.3 Average Building Heights New York City and Toronto as a Basis for Chap. 5

Table A.3 Average height of buildings in New York City and Toronto

City	Average height	Average total height
NYC	198	180
Toronto	162	

A.2 Luminance Values

The associated luminance photos are available online as additional material on the publisher's homepage for the book. All data and recordings were made by the author.

A.2.1 New York City

Table A.4 New York City luminance

Picture	Street	Sidewalk	Shop window + entrance	Facade
1 H NY	2		3.7	5.5
2 H NY	0.6	0.49	3.46	1.31
3 H NY	1.17	2.16	11.13	7.155
4 H NY		0.67	10.24	4.62
5 H NY	0.57	0.44	0.46	0.38333333
6 H NY	1.04	0.91	3.91	1.185
7 H NY	0.72	2.345	1.25	0.83
8 H NY	0.64	0.56	4.71	0.755
9 H NY	0.76	0.6		0.61
10 H NY	1.04	0.29	5.06	2.65
11 H NY		0.47	1.3	0.76
12 H NY	0.83	0.76	1.79	1.7
13 H NY	0.01236111	1.05	4.77	0.595
14 H NY	0.58			
15 H NY	1.54		4.49	1.19
16 H NY	0.53	0.55	2.4	0.555
17 H NY	0.44	0.635		0.765
18 H NY	4.16	1.3	10.34	1.02
19 H NY	1.2	3.39	3.58	0.735
20 H NY	0.67			0.4
21 H NY	0.44		10.58766667	1.57
22 H NY	1.4	2.28	7.485	2.35
23 H NY	0.69		5.35	1.19
24 H NY	1.06		11.09	5.57
25 H NY	0.82	0.77	3.33	0.8
26 H NY				2.59
Mean value	**0.99618961**	**1.09277778**	**5.258698413**	**1.87153333**
1 N NY		1.16		0.34666667
2 N NY	0.7	0.79		0.13
3 N NY	0.77	0.96		1
4 N NY	0.67	1.07		0.86
5 N NY	0.62	0.94		2.3
6 N NY	0.47	0.42		
7 N NY	0.44			
8 N NY	0.82			0.63
9 N NY	0.38	0.47		1.08
10 N NY	1.135	1.31		0.96
11 N NY	0.82	2.35		0.2
Mean value	**0.6825**	**1.05222222**		**0.83407407**

A.2.2 Boston

Table A.5 Boston luminance levels

Picture	Street	Sidewalk	Shop window + entrance	Facade
1 H W	1.62	2.285	4.89	2.95
2 H W	2.36	4.15	6.105	6.015
3 H B	1.89	1.055	1.97	1.78
4 H W	1.21	0.56		
5 H W	1.65	2.28	1.01	0.35
6 H W	1.34		3.22	1.21
7 H W	2.87		2.89	2
8 H W	1.27	1.19		0.58
9 H W	0.935	1.185		1.15
10 H W	0.44	1.51	4.64	1.9
11 H W	2.19	1.54	3.33	1.57
Mean value	1.61590909	1.75055556	3.506875	1.9505
1 N B	3.42	2.885		3.115
2 N B	3.3	2.58		1.465
3 N B	3.63	2.96		4.045
4 N B	0.115	0.29		1.12
5 N B	2.3			0.77
6 N B	0.95	1.32		
7 N B	0.13	0.17		0.25
8 N B	0.3	0.27		
9 N B	0.13	0.17		0.12
10 N B	1.05	0.99		0.77
11 N B	0.73	0.76		1.04
12 N B	1.66	1.51		0.59
Mean value	1.47625	1.26409091		1.3285

A.2.3 Toronto

Table A.6 Toronto luminance levels

Picture	Street	Sidewalk	Shop window + entrance	Facade
1 H D	0.78	0.56	1.03	1.31
2 H D	1.24	0.96	1.46	1.35
3 H D	1.33	6.32		1.25
4 H D	7.02	2.7		
5 H D	0.61	0.44	3.05	1.015
6 H D	1.23	1	3.555	1.9175
7 H D	0.98	1.87	4.8	4.15
8 H D	2.21	1.48	5.408	0.97
9 H D	0.62			
10 H D	1.01	1.26	1.415	0.925
11 H D	0.91		2.456666667	0.635
Mean value	**1.63090909**	**1.84333333**	**2.896833333**	**1.5025**
1 N T	1.16	0.68		
2 N T	1.66	1.07		0.975
3 N T	1.45	1.45333333		2.39
4 N T	0.63	1.04		0.715
5 N T	2.81	0.76		1.15
6 N T	2.18	0.86		1.555
7 N T	0.4	0.67		0.71
8 N T	0.97	1.12		1.435
9 N T	1.59	2.03		2.66
10 N T	2.77	1.115		2.765
Mean value	**1.562**	**1.07983333**		**1.595**

A.2.4 Warsaw

Table A.7 Warsaw luminance levels

Picture	Street	Sidewalk	Shop window + entrance	Facade
1 H W	1.29			2.48
2 H W	1.7		3.945	1.72
3 H W	4.6			4.79
4 H W	4.81		10.4	6.965
5 H W	3.79	3.92	17.19666667	6.215
6 H W	3.26		6.83	3.1
7 H W	5.27	10.42		
8 H W	7.76	7.62		
9 H W	7.29		8.92	5.38
10 H W	8.34			
11 H W		9.41	22.05	10.72
12 H W	4.58	6.85	13.31	7.78
13 H W	4.46	4.9		22.62
14 H W	17.14		202.3	67.84
15 H W		5.04	36.53	9.08
16 H W			15.37	6.17
17 H W	6.99			4.04
18 H W	7.51	4.93		4.47
Average value with 14 H W	**5.91933333**	**6.63625**	**33.68516667**	**10.8913333**
Average value without 14 H W	**5.11785714**		**16.82368333**	**6.82357143**
1 N W	7.47	4.11		6.38
2 N W	4.54	3.44		2.12
3 N W	3.16	1.68		3.18
4 N W	4.19			1.48
5 N W	6.3	3.95		
6 N W	2.42			7.62
7 N W	2.53			
8 N W	2.4			
9 N W	4.12			1.65
10 N W		2.34		2.94
11 N W	1.36			1.88
12 N W	4.3	3.09	2.8	2.565
13 N W	5.01		6.22	
14 N W	4.53	4.2	10.39	
15 N W	1.98		5.38	

(continued)

Table A.7 (continued)

Picture	Street	Sidewalk	Shop window + entrance	Facade
16 N W	1.74		16.67	12.96
17 N W		4.36		6.14
18 N W	3.99	2.89	18.9	
19 N W	3.34			3.705
Mean value	**3.72823529**	**3.34**	**10.06**	**4.385**

A.2.5 Hanover

Table A.8 Hanover luminance levels

Picture	Street	Sidewalk	Shop window + entrance	Facade
1 H H	0.36	0.1		
2 H H	0.49	0.22		0.21
3 H H	0.88	0.28	1.38	0.985
4 H H	0.65	0.35	0.2075	
5 H H	0.51	0.23		
6 H H	0.44	0.26	1.07	0.495
7 H H	0.42	0.23	0.73	0.19
8 H H	0.69	0.49		
9 H H	0.3	0.18	1.12	
10 H H	0.34	0.64		
11 H H	1.59		3.94	3.84
12 H H	1.17	0.9	1.14	0.74
13 H H	1.2	4.54	3.876666667	0.85
14 H H	0.79	1.15	0.97	0.395
15 H H	1.16	0.28		
16 H H	1.445	0.72	10.965	6.04
17 H H	0.42	1.31		0.4
18 H H		4.33	13.49	
19 H H		0.56	1.655	2.36
20 H H		1.18	13.93	4.9
21 H H		1.59	5.863333333	2.11
22 H H		0.24	1.37	0.455
23 H H	1.375	2.04	8.5	2.02
24 H H	0.855	0.51		0.27

(continued)

Table A.8 (continued)

Picture	Street	Sidewalk	Shop window + entrance	Facade
25 H H	0.67	0.22		0.29
26 H H		0.28	1.16	0.48
27 H H		0.23	1.24	0.61
28 H H				
Mean value	0.78775	0.298	1.407916667	1.144
1 N H	0.29			0.06
2 N H	0.28			0.22
3 N H	0.14	0.27		0.23
4 N H	0.27	0.2		0.2
5 N H	0.32			0.245
6 N H	0.5	0.63		0.13
7 N H	0.65	0.26		1.375
8 N H	0.34			0.16666667
9 N H	0.38	0.17		0.12
10 N H	0.22			
Mean value	0.339	0.306		0.30518519

Explanation of Terms

Expressions and Terms

Aerosols	Liquid and solid particles floating in the air.
Colour temperature	The colour temperature indicates whether the light source produces “cool” (high blue component) or “warm” (high red component) light. The color temperature is specified in Kelvin (see Philips Lighting Academy 2008).
Conventional light/lamp	All luminaire/lamp technologies, except LEDs.
LED Substitute	LED replacement for conventional lights.
Luminaire horizon	Imaginary horizontal line below the lightproof housing.

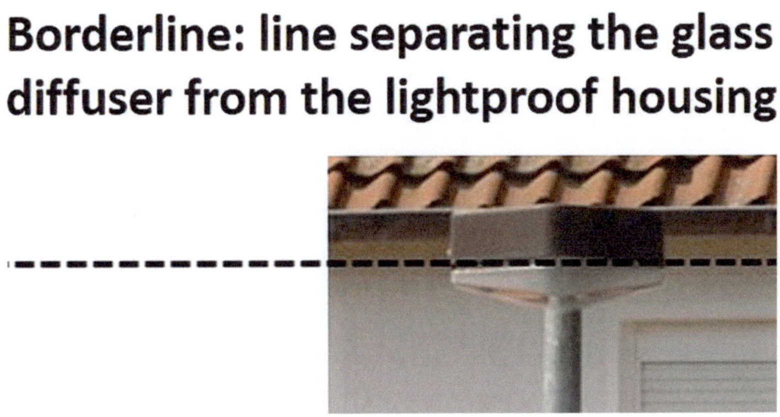

E. E. Goronczy, *Light Pollution in Metropolises*,
https://doi.org/10.1007/978-3-658-29723-7

Light point height	Term for the height of the light source.
Light column	Designation for a "rod-shaped" lamp (lamp model).
Light distribution curve	"The light ditribution curve indicates how much light is emitted by a luminaire at a certain angle. 0° is always vertical under the lamp." (translated from "Wir sind heller 2015")
Skybeamer	Powerful projector which is directed into the sky.

Photometric Data

Illustration 1 Luminous flux

Luminous flux Φ(Philips Lighting Academy2008)
This refers to the total light power emitted per second by a light source.
The unit of luminous flux is lumen (lm)
Examples:
- 75 W incandescent lamp: 900 lm
- 39 W fluorescent lamp: 3500 lm
- 250 W high-pressure sodium vapour lamp: 30,000 lm
- 2000 W metal halide lamp: 200,000 lm

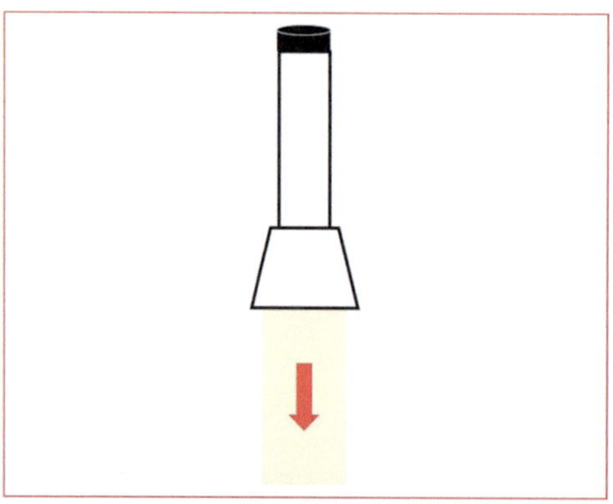

Illustration 2 Light intensity

Luminous intensity I
This is defined as the luminous flux emitted in a specific direction. The unit of luminous intensity is candela (cd).
Examples:
- 5 W bicycle lamp without reflector: 2.5 cd
- 5 W bicycle lamp with reflector: 250 cd
- 120 W reflector bulb: 10,000 cd
- Lighthouse: 2,000,000 cd

(continued)

Illustration 3 Illuminance

Illuminance E

This is the amount of light that falls on an area element of a surface. The unit of illuminance is lumen/m^2 or lux (lx)

Examples:

– Summer, with clear skies (equator): 100,000 lux
– Outdoors, under a cloudy sky: 5000 lux
– Artificial light, in a well lit office: 800 lux

Illustration 4 Luminance

luminance L

This refers to the light that is emitted by a surface element in a certain direction. The luminance is expressed in cd/m^2 (area viewed from the relevant direction).

Examples:

– Surface of the sun: 1,650,000,000 cd/m^2
– filament of a clear incandescent lamp: 7,000,000 cd/m^2
– Fluorescent lamp: 5000–15,000 cd/m^2
– road surface with artificial light: 0.5–2 cd/m^2

Structural Types of Lamps

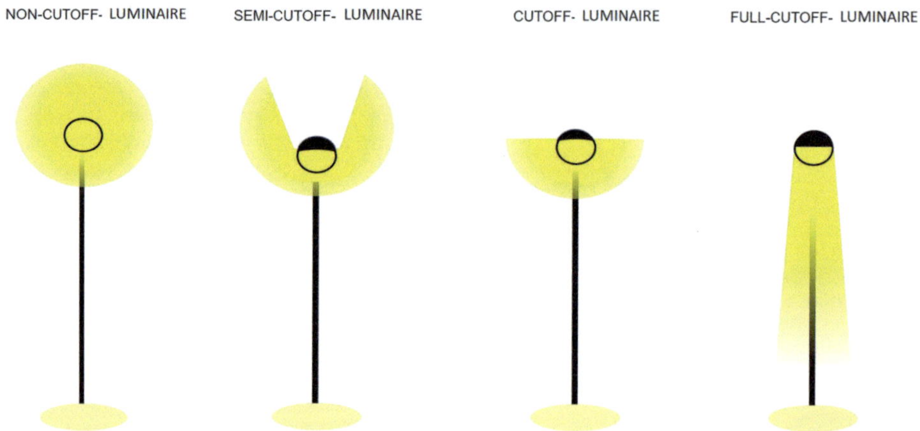

Illustration 5 Types of luminaire

MIX
Papier aus verantwortungsvollen Quellen
Paper from responsible sources
FSC® C105338

FSC
www.fsc.org

If you have any concerns about our products,
you can contact us on
ProductSafety@springernature.com

In case Publisher is established outside the EU,
the EU authorized representative is:
Springer Nature Customer Service Center GmbH
Europaplatz 3, 69115 Heidelberg, Germany

Printed by Libri Plureos GmbH
in Hamburg, Germany